Becoming

Beside

Ourselves

Becoming Beside Ourselves

The Alphabet,

Ghosts,

and

Distributed

Human Being

Brian Rotman

FOREWORD BY TIMOTHY LENOIR

DUKE UNIVERSITY PRESS

Durham & London

2008

Becoming Beside

Becoming is certainly not imitating, or identifying with something; neither is it regressing–progressing; neither is it corresponding, establishing corresponding relations; neither is it producing, producing a filiation or producing through filiation. Becoming is a verb with a consistency all its own; it does not reduce to, or lead back to, "appearing," "being," "equaling," or "producing."

—Gilles Deleuze and Félix Guattari, *A Thousand Plateaus: Capitalism and Schizophrenia*

Beside is an interesting preposition . . . because there's nothing very dualistic about it; a number of elements may lie alongside each other, though not an infinity of them. Beside permits a spacious agnosticism about several of the linear logics that enforce dualistic thinking: noncontradiction or the law of excluded middle, cause versus effect, subject versus object. . . . Beside comprises a wide range of desiring, identifying, representing, repelling, paralleling, differentiating, rivaling, leaning, twisting, mimicking, withdrawing, attracting, aggressing, warping, and other relations.

—Eve Kosofsky Sedgwick, *Touching Feeling: Affect, Pedagogy, Performativity*

Ourselves

But who is this "self" . . . and why is it at this particular juncture in the history of Western societies the very identity of the self becomes problematic?

—Raymond Barglow, *The Crisis of Self in the Age of Information: Computers, Dolphins, and Dreams*

All rights reserved. Printed and bound by CPI Group UK Ltd, Croydon, CR0 4YY
on acid-free paper ∞ *Designed by Amy Ruth Buchanan.*
Typeset in Carter & Cone Galliard by Tseng Information
Systems Inc. Library of Congress Cataloging-in-Publication
Data appear on the last printed page of this book.

Contents

Foreword

MACHINIC BODIES, GHOSTS, AND PARA-
SELVES: CONFRONTING THE SINGULARITY
WITH BRIAN ROTMAN

Timothy Lenoir

The specter of a postbiological and posthuman future has haunted cultural studies of technoscience and other disciplines for more than a decade. Concern (and in some quarters enthusiasm) that contemporary
technoscience is on a path leading beyond simple human biological im-
provements and prosthetic enhancements to a complete human makeover
has been sustained by the exponential growth in power and capability of
computer technology since the early 1990s. Also driving interest in such
futuristic scenarios has been the increasing centrality of computational
media to nearly every aspect of science, technology, medicine, and the
arts, combined with the digital communications revolution of the mid-
1990s spawning both the Internet and the rapid proliferation of mobile
computer-based communications that have already produced significant
changes in the organization and production of knowledge as well as in
the functioning of the global economy. The deeper fear is that somehow
digital code and computer-mediated communications are getting under
our skin, and in the process we are being transformed.

While limitations to silicon-based computing might have temporarily
deflated some of the more exotic predictions of futurists such as Ray Kurzweil or Hans Moravec, current developments connected with nanotechnology, quantum computing, biotechnology, and the cognitive neurosciences provide ample resources for sustaining and even encouraging their
posthuman imaginary. More than \$4 billion in government investments
worldwide in nanotechnology research and development by 2006 has produced some promising results: carbon nanotube wires have been developed

for ultra miniaturized electronics components; the first building blocks of a controllable computation in biological substrates at nanoscale have been achieved. In the next phase of the nanotech initiative, Mihail Roco, the senior advisor to the U.S. National Science Foundation and chief architect of the National Nanotechnology Initiative, predicts the development of active nanostructures that change their size, shape, conductivity, and other properties during use, enabling the production of electronic components such as transistors and amplifiers with adaptive functions reduced to single, complex molecules. By 2010 Roco predicts that researchers will cultivate expertise with systems of nanostructures, directing large numbers of intricate components to specified ends, including the guided self-assembly of nanoelectronic components into three-dimensional circuits and whole devices. Medicine could employ such systems to improve the tissue compatibility of implants, or to create scaffolds for tissue regeneration, or perhaps even to build artificial organs (Roco 2006, 39). In the fourth stage of the current nanotechnology initiative, after 2015–20, the field will expand to include molecular nanosystems: heterogeneous networks in which molecules and supramolecular structures serve as distinct devices. Among the products of this phase of development Roco predicts new types of interfaces linking people directly to electronics. When considered in light of current research successes in the development of brain–machine interfaces,[1] the sorts of scenarios envisaged by Kurzweil in recent texts such as *The Singularity is Near: When Humans Transcend Biology*, in which he charts the conditions for the merger of computer-based intelligence and human biology to occur around 2045, begin to sound eminently plausible (Kurzweil 2005, 138). While he does not endorse Kurzweil's notions of a futuristic singularity, Rodney Brooks sees a similar merger of (nanoscale) robotic technology with biotechnology on our horizon:

> We are on a path to changing our genome in profound ways. Not simple improvements toward ideal humans as is often feared. In reality, we will have the power to manipulate our own bodies in the way we currently manipulate the design of machines. We will have the keys to our own existence. There is no need to worry about mere robots taking over from us. We will be taking over from ourselves with manipulatable body plans and capabilities easily able to match that of any robot (2002, 236).

Brooks's admonition that we are machines on a continuous path of co-evolution with other machines prompts reflection on what we mean by "posthuman." If we are crossing to a new era of the posthuman, how have we gotten here? And how should we understand the process? What sorts of "selves" are imagined by Brooks and others as emerging out of this postbiological "human"?

Cultural theorists have addressed the topic of the posthuman singularity and how, if at all, humanity will cross that divide. Most scholars have focused on the rhetorical and discursive practices, the metaphors and narratives, the intermediation of scientific texts, science fiction, electronic texts, film, and other elements of the discursive field enabling the posthuman imaginary. While recognizing that posthumans, cyborgs, and other tropes are technological objects as well as discursive formations, the focus has been directed less toward analyzing the material systems and processes of the technologies and more toward the narratives and ideological discourses that empower them. We speak about machines and discourses "co-constituting" one another, but in practice, we tend to favor discursive formations as preceding, and to a certain extent breathing life into, our machines. The most far-reaching and sustained analysis of the problems has been offered by N. Katherine Hayles in *How We Became Posthuman* and her more recent book, *My Mother Was a Computer*. Hayles considers it possible that machines and humans may someday interpenetrate. But she rejects as highly problematic, and in any case not yet proven, that the universe is fundamentally digital, the notion that a Universal Computer generates reality, a claim that is important to the positions staked out by proponents of the posthuman singularity such as Harold Morowitz, Kurzweil, Stephen Wolfram, and Moravec. For the time being, Hayles argues, human consciousness and perception are essentially analog, and indeed, she argues, currently even the world of digital computation is sandwiched between analog inputs and outputs for human interpreters (Hayles 1999; 2005, especially 206–13). How we will become posthuman, Hayles argues, will be through interoperational feedback loops between our current mixed analog–digital reality and widening areas of digital processing. Metaphors, narratives, and other interpretive linguistic modes we use for human sense making of the world around us do the work of conditioning us to behave as if we and the world were digital. Language and ideological productions thus serve as kinds of virus vectors preparing the ground for the gradual shift in ontology. In the case of Wolfram and others, Hayles argues, the appropriation of computation as a cultural metaphor assumed to be physically true constitutes a framework in which new problems are constructed and judgments made. "On the global level, our narratives about virtual creatures can be considered devices that suture together the analog subjects we still are, as we move in the three-dimensional spaces in which our biological ancestors evolved, with the digital subjects we are becoming as we interact with virtual environments and digital technologies" (2005, 204). The narratives of the computational universe serve then as both means and metaphor. In our current analog/digital situation Hayles proposes an analytical strategy she calls intermediation to analyze the new processual human/machine texts of the posthuman era.

> As an embodied art form literature registers the impact of information in its materiality in the ways in which its physical characteristics are mobilized as resources to create meaning. This entanglement of bodies of texts and digital subjects is one manifestation of what I call intermediation, that is, complex transactions between bodies and texts as well as between different forms of media. Because making, storing, and transmitting imply technological functions, this mode of categorization insures that the different versions of the posthuman will be understood, in Kittlerian fashion, as effects of media. At the same time in my analysis of literary texts and especially in my focus on subjectivity, I also insist that media effects, to have meaning and significance, must be located within an embodied human world (2005, 7).

From the media-theoretic perspective Hayles adopts in *My Mother Was a Computer*—a perspective she refers to as Kittlerian—subjects are the effects of media. In order to make effective use of Hayles's theory of intermediation we need to understand how the complex transactions between bodies and our inscription practices might take place and how to understand the "entanglement" she describes of media with the formation of human subjects.

How can we think beyond the notion of virtual creatures as rhetorical devices and explore instead how the embodied human subject is being shaped by a technoscientific world? Can we get at the embodied levels of the interactive feedback loops Hayles describes to examine the metabolic pathways and emerging neural architectures through which these technologies are getting under our skin? Brian Rotman's latest book, *Becoming Beside Ourselves*, offers a solution to this problem in a profound and elegant analysis deriving from material semiotics. Rotman circumvents the issue of an apocalyptic end of the human and our replacement by a new form of *Robo Sapiens*. Instead he draws upon the work of anthropologists, philosophers, language theorists, and more recently cognitive scientists shaping the results of their researches into a powerful new argument for the co-evolution of humans and technics, specifically the technics of language and the material media of inscription practices. The general thrust of this line of thinking may best be captured in Andy Clark's phrase, "We have always been cyborgs" (2003). From the first "human singularity" to our present incarnation, human being has been shaped through a complicated co-evolutionary entanglement with language, technics, and communicational media. Drawing upon empirical findings from recent cognitive neuroscience, studies of sign language, gesture, and the impact of new imaging and computational sciences on contemporary habits of thinking, Rotman teases out the significance of Clark's apt phrase in a powerful framework of material semiotics. The materiality of media rather than their content is what matters. Communicational media are machines operating at the heart of subject formation. Like Gilles Deleuze and Félix Guattari, Rotman views machinic operations at the basis of consciousness and mind as an emergent phenomenon; and while unlike Wolfram (2002) and Edward Fredkin (2001), he is less certain about embracing cellular automata as the fundamental ontology of the universe, Rotman sketches out a position sympathetic toward Deleuze's and Guattari's notions of the human body being understood as an assemblage of mutating machines—a Body without Organs—rather than as a teleologically orchestrated organism with consciousness as the core of coherent subjectivity. Consistent with the flattening of differences between biological and inorganic machines central to contemporary nanotechnology and cognitive neuroscience, Deleuze and Guattari argued that, as bodies without organs, human assemblages are capable of absorbing a variety of entities, including other machines and organic matter. In Rotman's elaboration of this perspective, media machines are not just prosthetic extensions of the body, they are evolving assemblages capable of being absorbed into the body and reconfiguring the subject. Every medium, whether it be speech, alphabetic writing, or digital code, and each media ecology, such as the configuration of gramophone, film, and typewriter discussed by Friedrich Kittler (1992, 1999), projects a virtual user specific to it. This projected virtual user is a ghost effect: an abstract agency distinct from any particular embodied user, a variable capable of accommodating any particular user within the medium.

Materialist semiotics of the sort Rotman proposes in concert with recent work in cognitive neuroscience, studies of gesture and language from psychology and evolutionary ethology, and a variety of recent developments in the computational sciences may point the way.[2] The path Rotman pursues in addressing the questions of the subject, embodiment, and agency was suggestively if inadequately marked out by Deleuze and Guattari in their emphasis on the human as machinic assemblage and in Guattari's suggestive notion that techno-machines operate invisibly at the core of human subjectification, particularly what Guattari referred to as "a-signifying semiological dimensions (of subjectification) that trigger informational sign machines, and that function in parallel or independently of the fact that they produce and convey significations and denotations, and thus escape from strictly linguistic axiomatics" (1995, 4). For media philosophers the question is whether Deleuze's and Guattari's cryptic and sketchily developed theses about "a-signifying semiological dimensions" of subjectification can be put on a solid foundation of what might be called "corporeal axiomatics" in contrast to Guattari's reference to "linguistic axiomatics."

Several key ideas from Rotman's earlier books, *Signifying Nothing: The Semiotics of Zero*, and *Ad Infinitum . . . The Ghost in Turing's Machine: Taking God out of Mathematics and Putting the Body Back In. An Essay in Corporeal Semiotics*, help us appreciate the power of the new work, which in many ways forms a trilogy extending and completing the critical framework offered in those earlier projects. In order to appreciate Rotman's newest contribution, I find it important to have in mind his theses about the role of signifying systems in the constitution of subject positions and agency, powerful ideas that form the core of those two earlier books. The new work has enriched these frameworks from semiotics with consideration of the importance of "bottom-up" architectures and distributed modes of agency/subjectivity deriving from Rotman's considerations of recent work in the computational sciences, distributed cognition, and cognitive neuroscience.

In *Signifying Nothing, Ad Infinitum . . .* , and several related essays, Rotman engages in a deeply critical dialogue with recent work in the philosophy of mathematics, language, and philosophy of mind. He has crafted a semiotic approach to mathematics which builds on some suggestive fragments of Charles Sanders Peirce in discussing the relation of signs (in Rotman's case, mathematical signs), interpreting subjects, writing, and agency. He folds all this brilliantly with the work of Foucault and Derrida in fashioning an original semiotic approach to the principal questions of postmodern philosophy by examining as isomorphic signifying moves the near-simultaneous introduction into Western culture of zero in mathematics, the vanishing point in painting, and imaginary money in economic exchange. Rotman shows that in the shifts from Roman to Hindu numerals, from iconic to perspectival art, and from gold money to imaginary bank money a common meta-sign indicating the absence of other signs emerges from what Rotman calls a new set of semiotic capacities—public, culturally constituted, historically identifiable forms of utterance and reception which *codes* make available to individuals. This new meta-sign requires the formulation of a new sign-using agency, a secondary subjectivity, in order to be recognized: in mathematics the invention of algebra by Vièta; in painting the self-conscious image created by Vermeer and Velázquez; in the text, the invention of the autobiographical written self by Montaigne; in economics, the creation of paper money by gold merchants in London.

Rotman followed the astonishing *Signifying Nothing* with the equally brilliant *Ad Infinitum.* . . . Not for the fainthearted, the book pursues one of the defining ideologies of Western mathematics: namely, its characterization as pure and abstract disembodied reason. In many ways, this book can be read as the analog in the philosophy of mathematics and science studies to Derrida's attack in *Of Grammatology* on logocentrism and real presence as central to Western metaphysics. Building on the semiotic model he had developed in *Signifying Nothing* and an earlier article, "Toward a Semiotics of Mathematics," Rotman examines nineteenth- and early twentieth-century developments in semiotics, particularly the semiotics of Peirce and Saussure. He uses this framework to analyze the development of Hilbert's abstract mathematics and the problems relating to infinity in mathematical reasoning that were central to the creation of modern mathematical set theory by Frege, Gödel, Hilbert, and others. Similar to the argument in *Signifying Nothing*, Rotman shows that operations with mathematical signs, and particularly functions such as counting, differentiating, etc., cannot be undertaken without an implicit but never-acknowledged "mathematical agent"—the one who counts. This move, to acknowledge reason as always embodied, leads to a consideration of Turing's and Church's work on computability, which also depended heavily on assumptions drawn from the mathematics of infinite sets. For Rotman, "information" is not somehow immaterial, but deeply physical, so that computers no more than mathematicians can disregard the embodied character of reasoning.

Rotman's work on semiotics, embodied reason, and agency in these earlier texts provides a basic set of concepts and strategies for his analysis of gesture and the machinic body in *Becoming Beside Ourselves*. The revolutions in the computational sciences, the rise of cognitive science, and the explosive developments in recent neuroscience, particularly work on distributed cognition, the embodied mind, and emergence that have taken off since the 1990s, provide empirical sources for Rotman's discussion of ghostly subjects and machinic bodies. This incorporation of results and implications of empirical science is what really distinguishes Rotman's *Becoming Beside Ourselves* and allows him to elevate the discussion to an exciting new plane. Drawing upon a framework he originally set out in the 1988 essay, "Toward a Semiotics of Mathematics," Rotman elaborates in chapter 3, "Technologized Mathematics," one of his key theses about the ways in which computational devices anastomose with and reshape the human. In the intervening years since Rotman wrote *Signifying Nothing* and *Ad Infinitum* . . . , a revolution in the computational sciences took place, particularly through the developments of parallel computing—not even contemplated as relevant by Turing—and the emergence of techno-scientific fields (by which I mean fields in which the science is inextricably bound to the machines that enable it) of modeling, near to real-time simulation, visualization, and computer-generated virtual environments. Rotman discusses the impact of these developments in transforming conditions of proof in mathematics to a near-experimental discipline. He examines the role of images, diagrams, and other forms of mathematical representation, issues particularly relevant to recent discussions concerning the role of graphic methods, "imaging," and visualization in computational sciences. In this chapter Rotman analyzes in terms of his own distinctive brand of material semiotics the transformation being effected in our experience of bodies and of agency by computer-mediated communication and especially technologies of visualization, emerging virtual reality, and even haptic feedback in new digital systems for motion capture. He argues that these forms of representation are not just supplements to a linear, alphabetic text that carries the "real" meaning of an utterance, but constitute languages all their own.

Rotman's theory of signification aims at showing that each media regime and each system of signification projects a specific configuration of the subject and a horizon of agency as a consequence of its normal operation. Moreover, in Rotman's view these semiotic systems evolve with the media machines that embed them. They are techno-cultural artifacts that co-evolve with their human host-parasites. Conceived in this fashion language, media, and possibly the new generations of intelligent machines we imagine just over the horizon might be considered companion species dependent on, but also powerfully shaping, us through a co-evolutionary spiral. Indeed, offered as a replacement for what she regards as her now outdated earlier notion of the cyborg, Donna Haraway advocated a similar line of inquiry in her "Companion Species Manifesto": "Earth's beings are prehensile, opportunistic, ready to yoke unlikely partners into something new, something symbiogenetic," Haraway writes (2003, 32). Co-constitutive companion species and co-evolution are the rule, not the exception.[3]

Is there any foundation for relating this approach to the biological evolution of human cognition to the theory of signification and the notions of media machines discussed by Rotman, Kittler, Hayles, Hansen, and others? Rotman pursues this question deep into the structure of symbolic communication and its embodiment in the neural architecture of evolving human brains. He draws especially upon the works of Terrence Deacon and Merlin Donald on the evolution of language for considering the formative power of media technologies in shaping the human and some of the critical issues in current debates about posthumanity. For Deacon and for Donald what truly distinguishes humans from other anthropoids is the ability to make symbolic reference. This is their version of the Singularity; *Homo symbolicus*, the human singularity. Although language evolution in humans could not have happened without the tightly coupled evolution of physiological, anatomical, and neurological structures supporting speech, the crucial driver of these processes, according to Deacon, was *outside* the brain; namely, human cultural evolution. The first step across the symbolic threshold was most likely taken by an australopithecine with roughly the cognitive capabilities of a modern chimpanzee. Symbolic communication did not spontaneously emerge as a result of steady evolution in size and complexity of hominid brains. Rather, symbolic communication emerged as a solution to a cultural problem. To be sure the evolution of language could not have arisen without a primitive prerequisite level of organization and development of the neurological substrates that support it. But in Deacon's view those biological developments were more directly driven by the social and cultural pressures to regulate reproductive behavior in order to take advantage of hunting-provisioning strategies available to early stone-tool-using hominids. Deacon argues this required the establishment of alliances, promises and obligations linking reproductive pairs to social (kin) groups of which they were a part. Such relationships could not be handled by systems of animal calls, postures, and display behaviors available to apes and other animals but could only be regulated by symbolic means. A contract of this sort has no location in space, no physical form of any kind. It exists only as an idea shared among those committed to honoring and enforcing it. Without symbols, no matter how crude in their early incarnation, that referred publicly and unambiguously to certain abstract social relationships and their future extensions, including reciprocal obligations and prohibitions, hominids could not have taken advantage of the critical resources available to them as habitual hunters (1997, 401). In short, symbolic culture was a response to a reproductive problem that only symbols could solve: the imperative of representing a social contract. What was at stake here was not the creation of social behavior by the social contract as described by Rousseau, but rather the translation of social behavior into symbolic form.

Once the threshold to symbolic communication had been crossed natural selection shifted in dramatic ways. Deacon bases his model on James Mark Baldwin's original proposals for treating behavioral adaptation and modification as a co-evolutionary force that can affect regular Darwinian selection (Baldwin 1895, 219–23; 1902).[4] Baldwinian evolution treats learning and behavioral flexibility as a force amplifying and biasing natural selection by enabling individuals to modify the context of natural selection that affects their future offspring. Behavioral adaptations tend to precede and condition major biological changes in human evolution because they are more responsive than genetic changes. As Robert Richards points out, this is not a form of Lamarckism, since changes acquired during an organism's own lifetime are not passed on directly to offspring. Rather, Baldwin's model proposes that by adjusting behavior or physiological responses to novel conditions during the lifespan of the organism, an animal could produce irreversible changes in the adaptive context of future generations (Deacon 1997, 322–23).

Deacon uses Baldwinian evolution in a provocative way to address the question of the co-evolution of language and the brain. Though not itself alive and capable of reproduction, language, Deacon argues, should be regarded as an independent life form that colonizes and parasitizes human brains, using them to reproduce (1997, 436).[5] Although this is at best an analogy — the parasitic model being too extreme — it is useful to note that, while the information that constitutes a language is not an organized animate being, it is nonetheless capable of being an integrated adaptive entity evolving with respect to human hosts. This point becomes more salient when we think of language as a communication system and examine the effects of media, including electronic media, more broadly.

For Deacon, the most important feature of the adaptation of language to its host is that languages are social and cultural entities that have evolved with respect to the forces of selection imposed by human users. Probably the primary selective force on language evolution is the pressure to produce linguistic operations that children can learn quickly and easily. Just like computer interfaces, languages evolve around user-friendliness, and languages that do not adapt to their user-niche disappear. In this Baldwinian co-evolution of physical and cultural processes, neurological structure and language ability — brain and speech — interact: changes in the newborn's brain, manifest in its cognitive abilities, exert a selective pressure on which features of language are learnable and which are not; this in turn feeds back, influencing the brain to change in certain ways, which further impacts the grammatical structures and semantic possibilities of spoken language available to the infant. And so it goes. The outcome, over thousands of generations, is the emergence of human language in which symbolic reference emerges from and is distributed among a web of evolutionarily older forms of reference. While apes, chimpanzees, and other animals are able to operate with symbols in certain carefully constructed contexts, everyday human cognition demands the construction of novel symbolic relationships. A considerable amount of normal, everyday problem solving involves symbolic analysis or efforts to figure out some obscure symbolic association. Deacon argues that the greater computational demands of symbol use launched selection pressure on increased prefrontalization, more efficient articulatory and auditory capacities, and a suite of ancillary capacities and predispositions which eased the new tools of communication and thought. Each assimilated change added to the selection pressures that led to the restructuring of hominid brains.

More than any other group of species, hominids' behavioral adaptations have determined the course of their physical evolution, rather than vice versa. Stone and symbolic tools, which were initially acquired with the aid of flexible ape-learning abilities, ultimately turned the tables on their users and forced them to adapt to a new niche opened by these technologies. Rather than being just useful tricks, these behavioral prostheses for obtaining food and organizing social behaviors became indispensable elements in a new adaptive complex. The origin of "humanness" can be defined as that point in our evolution where these tools became the principal source of selection on our bodies and brains. It is the diagnostic trait of *Homo symbolicus* (Deacon 1997, 345).[6]

In Deacon's theory evolutionary selection on the prefrontal cortex was crucial in bringing about the construction of the distributed mnemonic architecture that supports learning and analysis of higher-order associative relationships constitutive of symbolic reference. The marked increase in brain size over apes and the beginnings of a stone tool record are the fossil remnant effects of the beginnings of symbol use. Stone tools and symbols were the architects of the *Australopithecus-Homo* transition and not its consequences.

Symbolic reference is not only the source of human singularity. It is also the source of subject formation in all its varied manifestations. Directly relevant to Rotman's discussion of modern subject formation in distributed computer networks is the hierarchical structure of reference central to Deacon's and Donald's work. Like Rotman, Deacon bases his theory of reference on Peirce's semiotics. Peirce made the distinction between iconic, indexical, and symbolic forms of reference; where icons are mediated by similarity between sign and object, indices are mediated by some physical or temporal connection between sign and object, and symbols are composed of relations between indices and mediated by formal or conventional links rather than by more direct neurological connection between sign and object.

For both Rotman and Deacon symbolic reference is virtual, unreal, and tarries with the ghostly. Symbolic reference rests on the powerful combinatorial, associative logics of forming relationships between signs, and its mnemonic supports need only be cashed in and reconstructed in terms of their lower level indexical and iconic supports when needed. Symbolic reference is so powerful because it allows us to ignore most of the vast web of word-object, word-word, and object-object indexical associations and make rapid calculations using the mnemonic shortcut of symbol-symbol relationships instead. It is this virtual character of symbolic reference that is the source of its power and of its interest for our concerns with subject formation. The ignored indexical relationships are still the implicit grounding of word reference, but these interpretive steps can be put off until it can be determined exactly which are relevant and which are not. For Deacon symbols are neurological tokens. Like buoys indicating an otherwise invisible best course, they mark a specific associative path, by following which we reconstruct the implicit symbolic reference. The symbolic reference emerges from a pattern of virtual links between such tokens, which constitute a sort of parallel realm of associations to those that link these tokens to real sensorimotor experiences and possibilities. Thus it does not make sense to think of the symbols as located anywhere within the brain, because they are relationships between tokens, not the tokens themselves; and even though specific neural connections may underlie these relationships, the symbolic function is not even constituted by a specific association but by the virtual set of associations that are partially sampled in any one instance. Widely distributed neural systems must contribute in a coordinated fashion to create and interpret symbolic relationships. It is this virtual aspect of symbolic reference that leads to some interesting possibilities and peculiarities of subject formation. As Lao Tsu wrote, "Thirty spokes share the wheel's hub, but it is the hole in the center that provides its usefulness" (*Tao Te Ching*, quoted in Deacon 1997, 433).

All of this has important consequences for consciousness and subject formation. Three points are especially relevant to Rotman's discussion. The power of symbolic reference is due to its *virtual* character; its *shared* deployment; and its exteriority, the fact that it is largely *external* to the individual mind, being located in cultural systems and artifacts. From an evolutionary perspective, consciousness is always consciousness *of* something, and hence involves some form of representation. Animal minds, even of minimal complexity, construct and process internally generated indices with respect to an external world to which they are partially adapted. All nervous systems irrespective of their size and complexity support iconic and indexical representational processes. These are the basic ingredients of adaptation. Human brains share a design logic with other vertebrate brains, and so we share those aspects of consciousness that are mediated by iconic and indexical representations that those other species experience.

But minds capable of symbolic representation operate on a radically different playing field. Symbolic representation is not just a further complex addition to the capabilities of animal brains. Rather it has completely made over the human brain to aid language processing. As a result humans, capable of symbolic reference, are able to form an independent mental representation of another. Being confined to indexical representations animal brains can represent associations between stimuli, including the behaviors of others, and these relationships can be complex extending to a familiarity with the predispositions of others. But the step of forming an independent mental representation of the subjective experience of another requires an abstraction only possible with symbolic reference. Indeed, our ability to inhabit the different personas constructed by writers is only made possible by the structure of symbolic representation. Symbolic representation maintains reference irrespective of indexical attachment to any particular experiences, so that when a narrative of someone's experience is reconstructed by another, it can be regrounded in terms of the reader's or listener's own experience by interpreting it in terms of the iconic and indexical representations that constitute the listener's memory. Symbolic reference is interpreter-independent; it strips away any necessary link to the personal experiences and musings that ultimately support it. Although all readers of a novel share a common symbolic understanding of the events narrated, each individual's experience in response to them is distinct.

Unlike the interpretation of icons and indices (a process uniquely personal and insular within each brain), symbolic representations are in part externally interpreted. They are shared. Symbolic reference is at once a function of the whole web of referential relationships and of the whole network of users extended in space and time. It is as though the symbolic power of words is only on loan to its users. If symbols ultimately derive their representational power, not from the individual, but from a particular society at a particular time, then a person's symbolic experience of consciousness is to some extent society-dependent—it is borrowed. Its origin is not within the head.

> Consciousness of self in this way implicitly includes consciousness of other selves, and other consciousnesses can only be represented through the virtual reference created by symbols. The self that is the source of one's experience and intentionality, the self that is judged by itself as well as by others for its moral choices, the self that worries about its impending departure from the world, this self is a symbolic self. It is a final irony that it is the virtual not actual reference that symbols provide, which gives rise to this experience of self. This most undeniably real experience is a virtual reality (Deacon 1997, 452).

The theories of cognitive evolution upon which Rotman draws all point to the extraordinary flexibility of the neurological architecture of human cognition. Humans as a species have plastic, highly conscious nervous systems, the capacities of which allow us to adapt to intricate challenges of a changing cognitive environment. Rather than being rigidly hard-wired to structures inside the brain, symbolic communication created a mode of extrabiological inheritance with a powerful and complex character, and with an autonomous life of its own. The individual mind is a hybrid product, partly biological and partly ecological in origin, shaped by a distributed external network the properties of which are constantly changing.

The work we have discussed by evolutionary biologists and cognitive neuroscientists has dealt primarily with the origins of language, particularly with speech and to a limited extent with writing. This work has emphasized that the leap to the symbolizing mind did not depend on a built-in hard-wired tendency to symbolize reality. The direction of flow was *from* culture to the individual mind, from *outside-to-inside*.[7] Rotman is interested in expanding this analysis to include media other than speech and writing, especially technologically mediated and computer-based forms of communication. He locates rich suggestions for the path to pursue in the work of Deleuze and Guattari.

It is in chapter 2 on gesture that Rotman's "Deleuzian turn" takes flight most dramatically. A number of "new humanists" are interested in drawing upon work in the natural sciences, particularly recent work in cognitive science and fields such as functional MRI, to shed new light on traditional humanistic questions, such as the role of emotion and affect in processes of reasoning, previously treated as the territory of abstract, disembodied mind. Other areas in which the findings of empirical science are impinging on recent work in the humanities are analyses of neuroaesthetic bases for artists' construction and viewers' appreciation of images and analysis of the role of narrative and metaphor in normal reasoning processes. But Rotman's use of these recent scientific fields is the most impressive—in fact, the only—attempt with which I am familiar of an analysis aimed at resolving key issues in postmodern philosophy. In chapter 3 on computing and mathematics and in chapter 1 on the "Alphabetic Body" Rotman draws on the work of recent cognitive neuroscience to advance his "Deleuzian" theses on the machinic body as an assemblage of relatively independent and autonomous units. For instance, recent work in the neurosciences supports a reevaluation of the human innate sense of number. Rather than being associated with a singular "faculty," particular organ, or region of the brain, counting, according to the cognitive neuroscientist Stanislas

Dehaene, is not even an evolved skill, but a capacity assembled from different and independent brain activities each on their own having nothing to do with number.

What Rotman offers in *Becoming Beside Ourselves* is an extension and revitalization of Derrida's grammatological project, updating it for the computationally intensive environments that increasingly constitute us as posthuman. In his critique of Western metaphysics Derrida pointed to the logic of the supplement as a way of deconstructing the power of logocentrism. Whereas speech is normally treated as primary, the site of presence to oneself, and writing is treated as secondary, a supplement that comes after speech as a technique for notating it, Derrida argued that writing comes *before* speech. Writing, Derrida argued, is really a form of graphism, a general mode of signifying operating in all cultural production. Writing and speech are both dependent on this higher form of writing, what Derrida called arche writing, a writing before the letter.

But Derrida did not successfully extend this provocative analysis to other signifying systems such as images. As Rotman points out, Derrida's commentary on Husserl's *The Origins of Geometry*, for instance, never mentions the complete absence of diagrams or reference to them throughout the text. While critiquing the Western metaphysical system, Derrida's deconstruction remained bound to the world of print it called into question. Rotman's project on corporeal writing introduces the exploration of graphism in the wider sense as a synthesis of semiotics, computation, and experimental science.

Rotman's brilliant treatment of gesture, speech, and their relations to other signifying systems moves consideration of the posthuman subject onto a new page of clarity and rigor. Some of the resources Rotman draws upon are neurological studies that argue for a form of internal touching, a virtual auto-hapticity that appears to be the condition for self-consciousness. He draws on empirical investigations of verbal narration in cognitive psychology that demonstrate that gesticulation—previously believed to be an unnecessary supplement to speech and thought—is a deep component of utterance, having to do with the semantics, pragmatics, and discursive aspects of speech. These studies call into question the causal or conceptual priority of speech by demonstrating a tight temporal binding, accurate to fractions of a second, discovered to operate between gesticulation and speech. Drawing on these sorts of empirical studies, Rotman argues that thinking that eventuates in speech has its origins in pre-verbal visuo-kinetic images which then become gesticulated and verbalized to form utterance. Spoken thought starts as a yet-to-be realized gesture. Thought, including abstract thought such as mathematical reasoning, rests on metaphors and diagrams derived from repeated and deeply layered patterns of body movement. Moreover, he draws upon other work on phonetics and speech synthesis which demonstrates that it is precisely as a gestural system that speech is best apprehended and perceived by a listener. Drawing on the work of the evolutionary neurobiologist Terrence Deacon once again, Rotman argues that auditory processing of speech sounds does not appear to be based on extracting basic acoustic parameters of the signal before mapping them onto speech sounds. Speech sounds appear designed instead to predict which oral-vocal movements produced them and ignore the rest. We listen, Rotman concludes, not to speech sounds as such but to what they signal about the movements of the body causing them. We listen to speech as symptoms of gestures.

In two very engaging chapters, "The Alphabetic Body" referred to above and "Gesture and Non-Alphabetic Writing," which introduces the notion of the body without organs of speech, Rotman explores recent work on American Sign and argues for gesture and other non-alphabetic systems of expression as languages in their own right. In these chapters, Rotman points to the importance of the materiality of media in "wiring" thought, and in the chapter on alphabetism, even draws upon work from ethology and neurobiology to bolster his claim that in the historical process of establishing non-pictorial alphabetism as the basis for thought, the alphabet disrupted "the integrated complementarity of upper and lower, tongue and larynx, articulation and breath, consonant and vowel; it effects a pulling apart and deactivating of the circuits between neocortex and the midbrain." (Rotman 2002, 97) The powerful result of this disjunction, he argues, was the hierarchical subordination of the midbrain to the neocortex, and attendant to that the cutting loose of words from voice. Going one step further, Rotman argues that the result of this vertical neurological separation was the institution of a primary dualism of mind over body, the ultimate expression of which is the notion of the pure, disembodied mind, that constant theme threaded through all Rotman's work.

What then about the posthuman? Are we transitioning to some new form of self adapted to our environment of ubiquitous computing technology, and if so, how is this self assembled and transformed by the machinic processes of our technoscientific milieu? Since the rise of *Homo sapiens* between 100,000 and 200,000 years ago, there has been little change in brain size or, as far as can be determined, in brain structure. A critical contributing factor to the rapid cultural evolution that took off with *H. sapiens* and has continued at an ever-increasing pace since is the development of supplements to individual internal biological memory in the form of visuo-graphic systems and external memory media, especially written records and other forms of symbolic storage (see especially Donald 1991, 308–12). Rather than being limited by our neural architecture, these external material supports have only enhanced the symbolizing power of the mind. In a sense, the recent development of the Internet and distributed forms of electronic communication only further accelerate a process that has defined and shaped human being since that first singularity. From the perspective of the work in evolutionary cognitive science we have discussed, any change in the way information gets processed and represented inevitably constitutes a change in the cognitive economy of the subject, a difference in psychic architecture and ultimately of consciousness itself. Teasing out the implications of this notion, Rotman argues that the medium of alphabetic writing introduced as silent collateral machinic effects an entire neurological apparatus enabling practices, routines, patterns of movement and gestures, and kinematic, dynamic, and perceptual activities as part of the background conditions—in terms of Deleuze and Guattari, the a-signifying dimensions of the medium lying beneath the medium's radar as part of its unconscious—giving rise to the lettered self, a privately enclosed, inward and interiorized mind, structured by the linear protocols and cognitive processing that reading and writing demand.

This mode of psychic organization is giving way to new forms as part of the massive shift in computational media taking place. Of particular importance to our present situation is the influx of parallel computation into what has been almost exclusively a serial computational regime. The parallel/serial duo is nothing new. In fact, as Rotman argues, the dynamic tension between parallel and serial modes of thinking and representation have characterized every media regime. Rotman examines the tandem opposition of serial and parallel forms across many types of activity: music (melody versus harmony), symbolic forms (text versus image), arithmetic (ordinal versus cardinal numbers), film editing (Eisenstein versus intercut montage), electrical circuits (serial versus parallel), and especially serial versus parallel modes of computing. The serial/parallel duo come together and are always in a certain tension with one another. Seriality is exemplified in narratives, routines, algorithms, melodies, timelines; parallelism is exemplified in scenes, episodes, harmonies, contexts, atmospheres, and images. Parallelism foregrounds presence, simultaneity, co-occurrence. Serialism foregrounds linear order and sequence and occurs in counting, listing, lining up, and telling. Serialism privileges a certain mode of cognitive and psychic organization, according to Rotman: namely, the individual mind/brain in which thinking takes place inside the closed, individual thinker. Everything outside the individual symbol-processing brain is assigned to context and plays no substantive role in the thinking process. This model of the mind and of thinking is being challenged and displaced by the researches of contemporary cognitive science, which are demonstrating that what was previously marginalized as context is actually a crucial element in how we think. Not only is thinking always social, culturally situated, and technologically mediated, but individual cognition requires symbiosis with cognitive collectivities and external memory systems to happen in the first place. Parallel computing, Rotman writes, puts into flux the relations between internal self and external other, "since it is a machinic implementation, not of individual linear thinking but of distributed bio-social phenomena, of collective thought processes and enunciations, that cannot be articulated solely on the level of an isolated, individual self. Its effects are to introduce into thought, into the self, into the 'I' that engages its various forms, parallelist behavior, knowledge, and agency that complicate and ultimately dissolve the idea of a monoidal self" (92).

Long characterized by linear processing of code, computing is undergoing a massive shift toward parallelism. In nearly every venue of computing, from high-end processing of massive data sets, such as the human genome, and large-scale imaging projects, such as GIS maps, to routine gaming machines such as the Sony PlayStation 3, computing is being performed by multiple machines working simultaneously in parallel on different parts of the job to be computed, or (as in video-editing workstations, game machines, and even some new laptops) multiple processors in the same machine. In addition the computational affordances of cell phones, pervasive technologies for multi-tasking such as instant messaging, manipulation of multiple avatars of the self in communally inhabited virtual worlds such as *World of Warcraft* and *Second Life*, and engagement with a variety of forms of distributed agency, blends of artificial and human agents in networked circuits—all these contribute, Rotman argues, to making the parallelist self radically different from the single, serial, alpha-beticized psyche it is in the process of displacing.

Both crucial to and symptomatic of this shift to parallelism is the centrality of visualization technology and of the strategic influx of images into all forms of contemporary cognitive work. Everywhere pragmatic images, graphs, charts, tables, figures, maps, simulations, and other forms of visual artifact are permeating our reading and writing practices. These apparently innocuous information-bearing, instructional, explicatory, and otherwise instrumentally oriented images are, from Rotman's perspective, a (welcome) dimension of parallelism, prompting him to cite artist Helen Chadwick's dreamlike meditation, "What if dangerous fluids were to spill out, displacing logic, refuting a coherent narrative, into a landscape on the brink of I?" Nothing better represents this "spillage of the Ego" as a prelude to the emergence of a para-self, Rotman urges, than the prevalence of the post-photographic digital image, and especially the GIS map. The post-photographic image dissolves the classic viewer rooted in Renaissance perspectivalism that privileged a self with a point of view outside the imaged object. An increasingly familiar example of what Rotman is describing occurs in our obsession with GIS maps, such as maps provided by Google Earth, with multiple (currently up to twelve) separate graphic layers overlaying different kinds of information that can be dynamically viewed as a co-present assemblage of images and proactively navigated by the user. GIS maps of this sort enact parallel seeing of images that previously had to be viewed side by side, serially; in the process they reshape the fixity of the viewing subject and promote a dynamic viewing body that bypasses a perspectival mode of viewing. In terms of Rotman's thesis, this dissolution of seriality impinges directly on the subject and the construction of the self, a falling away from a one-dimensional, singular consciousness into parallel, distributed co-presence. Rotman summarizes this transition eloquently:

> Once, not so long ago, there was an absolute opposition of self and other: an 'I,' identical to itself, wholly present as an autonomous, indivisible, interior psyche against an external, amorphous collectivity of third persons outside the skin. Now the I/me-unit is dissolving, the one who says or who writes 'I' is no longer a singular integrated whole, but multiple: a shifting plurality of distributed I-parts, I-roles, I-functions, and I-presences. Now the 'I' bleeds outward into the collective, and the collective introjects, insinuates, and internalizes itself within the me.

What was privately interior and individual is invaded by the public, the historical, the social. What was the outside world of events enters (and reveals itself as having always entered) the individual soul in the mode of personal destiny (99).

As we spend more time in electronically mediated environments, engaging with massively parallel distributed computing processes that are merging ever more seamlessly with the material processes and technological affordances of our everyday world, we are, in Rotman's terms, becoming, literally evolving, as distributed machinic multiples, para-selves beside our selves.

Preface

Becoming Beside Ourselves: The Alphabet, Ghosts, and Distributed Human Being is the third in a trilogy devoted to the nature and functioning of certain signs and the writing practices associated with them. In each essay I explore ideas, imaginings, conceptual innovations, subjectivities, and forms of consciousness which the signs facilitate (and prevent), as well as the absent agencies and metaphysical beliefs that seem to arrive with them. The first essay, *Signifying Nothing: The Semiotics of Zero*, focused on the mathematical zero, 0—its contested entry into European thought as an infidel and paradoxical concept; its relation to various understandings of 'nothing' and emptiness; its disruptive semiotic effects; its midwifing of cognate forms in painting (the vanishing point) and economics (imaginary money); its secondary formations; and the complexities of its dual mathematical role: as a number like any other and as a meta-sign at the heart of the familiar Hindu-Arabic place notation that assigns names to the endless progression 1, 2, . . 10, 11 . . . of whole numbers. This latter role, essentially a marking of absence, serves also to signify the mathematical origin (of counting, of a system of coordinates) by standing in for an absent person, indicating the trace but not the presence within mathematical texts of the one-who-counts or coordinatizes, and thereby functioning as a sign of the "necessary residue" within mathematics, as Hermann Weyl put it, "of ego extinction."

The second essay, *Ad Infinitum . . . The Ghost in Turing's Machine: Taking God out of Mathematics and Putting the Body Back In. An Essay in Corporeal Semiotics*, pursues the metaphysical and disembodied agencies that accompany the writing and thinking of the mathematical infinite. A concept one encounters immediately in the diagram '. . .' of endless continuation ('potential' infinity), and in the ideogram '∞,' the mathematical inverse ⅟₀ of zero, comprehending a 'completed' infinity of passage to the limit.

From classical until recent times (and for some still today) these two infinities were opposed and not equally meaningful. For Aristotle only the former, potential, infinite—the ability to always add another unit and hence to count without end—made sense; an ability intrinsic to his conception of 'mind,' or *nous*, the incorporeal organ of thought, and he rejected the actual infinite as the source of the infamous paradoxes of Zeno. With the entry of zero-based place notation in the Renaissance, the actual infinite entered mathematics in the form of a completed infinite sum ($1 = \frac{1}{2} + \frac{1}{4} + \frac{1}{8} + \ldots$), and, then, in a more radical form toward the end of the nineteenth century, when Georg Cantor introduced transfinite numbers and a hierarchy of infinities. Echoing Aristotle, the constructivist mathematician Leopold Kronecker rejected the reality of such abstractions, condemning them as artificial and unreal compared to the 'natural,' potential infinity enshrined in the progression 1, 2, 3, . . . of whole numbers: "God," he insisted in a remark widely attributed to him, "made the integers, the rest is the work of Mankind."[1]

But what is it that makes the numbers 'natural,' that prompts mathematicians to call them so and conceive of them as given, always somehow 'there,' before and independent of the human mind and its works? And why invoke God as their creator? In a sense such questions seek to out-Kronecker Kronecker by insisting that even the potential infinity of numbers is a cultural construct, the "work of Mankind." This leads to a questioning of Aristotle's *nous*, the source of the supposedly natural ability to repeat endlessly. On the contrary, the ability requires a supernatural—disembodied—agency operating, as God is presumed to operate, metaphysically, outside the exigencies of time, space, energy, and physical presence; in short, an immaterial ghost. Some such agency is invoked by mathematicians (though they do not describe it so) when they write '. . .' and imagine the endless continuation of numbers signified by it. To say more, one can try to imagine what it would be like to conceive counting and think numbers outside a metaphysics of the infinite. One can ask what it might mean to iterate in *this* universe, the one which in-corporates us, which we embodied humans in-habit, in contrast to the transcendent ghost-space appropriate to a bodiless 'mind.' Under the rubric of non-Euclidean counting I sketch one idea of what this might involve.

Becoming Beside Ourselves pushes the question of supernatural agencies further back, situating them mediologically, in the context of their production. It asks: How, by what material, cognitive-affective, means, did God and 'mind' come to be—to exist, to be known, to be objects of belief—within Western culture? Unlike its predecessors, this essay's primary orientation is not mathematical signs. True, God and Mind arrive floating ghostlike inside the mathematical infinite; their *origins*, however, are in the letters of the alphabet which write speech, not in mathematical ideograms and diagrams inscribing ideas. Specifically, I argue the following thesis. God and 'mind' (Mind, *nous, psyche*, soul) are media effects of the alphabet, hypostatized entities, ghosts that emerged from the writing of 'I' in the sixth century BCE within the respective Jewish and Greek deployments of alphabetic writing, born at a point when the medium had become naturalized, the effects of written mediation invisible.

Crucial to the argument is a fundamental mediological point, namely the insistence that any act of self-enunciation is medium specific. This immediately implies that the four reflexive acts—a gestural self-pointing 'I'; an 'I' spoken in language; an inscribed 'I' within alphabetic writing; and a digital 'I,' a self-enunciation within contemporary network media— though interconnected and co-present are to be distinguished from each other. They project different sorts of affect, have different relations to embodiment, operate differently in their milieus, and engender their own forms of subjectivity. In particular, and of cardinal importance for the existence and features of God and Mind, to utter 'I' and to write 'I,' despite their everyday conflation within Western textual discourse, are radically different signifying acts; and it is precisely the fusing of them, the near universal unawareness of their difference and what might turn on it, that provides the rhetorical matrix for belief in disembodied agencies known only through writing.

Along with belief in their existence are their profiles, the features their believers attribute to them, and coupled to this the *affect* these agencies project. The affect proper to human speech, which pertains to moods, feelings, passions, attitudes, or emotions it conveys and induces, lies in its *tone*, a phenomenon determined by the gestures of the voice, those auditory movements of the body within utterance: its hesitations, silences, emphases, sharpness, timbre, musicality, changes of pitch, and other elements of prosody. The alphabet knows nothing of all this. It eliminates tone and any kind of prosody completely: it reduces the voice to words and writes 'what's said' but not the manner of its saying, its delivery, how what's said is said. What, one can ask, would be the features of a 'speaker,' and the affect of a 'voice' known only through alphabetic writing?

Since tone is the presence and action of the body in speech, such a figure would be disincarnate and invisible, as indeed both God and Mind are. But tone of voice also serves two other functions: it is the means by which a speaker registers the presence of and attitude toward a listener, and it is an important means—one of the earliest—humans have of individuating themselves and others. Lacking all tone, an agency known only through the alphabetic inscription of its words would appear abstract and (chillingly) indifferent to the existence or not of its supposed addressee; it would also, by being unlocatable and unspecifiable as an individual, project an 'other'—nonhuman—form of identity.

But the alphabet's twenty-five hundred year textual domination of Western culture, "the era of alphabetic graphism" (Leroi-Gourhan 1993), with its singularizing, monadic, and linear logic appears to be ending. All that was founded and so long held in place by the alphabet's mediation becomes increasingly difficult to sustain confronted with the parallelist and distributed logic of multiplicities. In this light, once revealed as media-effects, God and Mind along with 'soul' become no longer tenable items of belief and begin to feel strange and of diminishing relevance within the increasingly networked contemporary scene, the lettered self which co-evolved with these agencies and gave them credence now past its heyday and increasingly overshadowed by a new self-enunciation. This fourth 'I' which is beginning to disrupt and reconfigure its gestural, speaking, and writing predecessors, is a plural self, a self beside itself, which cannot but appear as unstable, virtual and 'unreal' with respect to us—the deeply embedded denizens of alphabetic culture.

Acknowledgments

My thanks to the many audience members at meetings of the Society for Science and Literature for providing a sympathetic forum for several of the ideas here. My thanks to friends and colleagues in the Department of Comparative Studies at The Ohio State University, whose rare multidisciplinary make-up and comparativist ethos have made a perfect place to develop ideas that are off conventional academic maps. My thanks to Tim Lenoir for the generous support and critical evaluation he has shown my work over the many years we've known each other. My gratitude to the fertile philosophical writings of Gilles Deleuze on multiplicities, becomings, Spinoza, and much else, and to the imaginative reach of Félix Guattari's thinking of technology "at the heart of the subject" which permeate this essay. And my thanks to the friendship, collaboration, and wit of Rich Doyle for the affect, the enlightenment, and the prodigious bouts of laughter we shared over the course of hundreds of telephone conversations.

My thanks to the staff of Duke University Press. To the editorial director, Ken Wissoker, for taking on the project and for his critical promptings that improved the manuscript; to my editor Courtney Berger whose suggestions, guidance, and tolerance of last-minute changes are much appreciated; to Amy Ruth Buchanan for her elegant design; and to Mark Mastromarino for his meticulous and always helpful copyediting.

Thanks to the Royal Society for permission to reprint "Will the Digital Computer Transform Classical Mathematics" that appeared in *Philosophical Transactions, Royal Society*, London, Series A, 361 (2003): 1675–90; to Johns Hopkins University Press and the Society for Science and Literature for permission to reprint "Corporeal or Gesturo-Haptic Writing" that appeared in *Configurations* 10 (2002): 423–38; to the University of Wisconsin Press, for permission to reprint "Going Parallel" that appeared in *SubStance* 91 (2000): 56–79; to Taylor and Francis for permission to reprint "The Alphabetic Body" which appeared in *Parallax* 8 (2002): 92–104; and to the University of Memphis for permission to reprint "Aura" that appeared in *River City* 20 (2000): 14.

Aura

It was as if everything she thought, felt, remembered, had an aura; behind the briefest eye-blink, the least flicker of touch, a shadow, a kind of ghost presence. This aura wasn't part of her, nor was it in any sense that she could fathom produced by her, nor did it seem answerable to her, even though—of this she was certain—only she was aware of it. Neither did it seem to precede her and her business: it wasn't there before—couldn't be anywhere—before she felt what she was able to feel, before she perceived what she—her body—decided she could perceive. And yet there it was, ghost of things present, faint pulsation of the real; at times like the glow on the surface of the universe, at others the dark outline of a world dazzled by there being nothing in it but its own presence. Often it was neither light, nor dark, nor anything visible, but just a presence— simply there—clinging to the motion of her being in space—like the field of a magnet, or radiation from the earth's rocks. Of late it had occurred to her: perhaps it was the aura that was real, felt things, had a body, sat and moved through space and perceived the countless pulsations of light and energy in the universe, and that she was the shadow clinging to it, following it around, copying its business before it had time to look around and be aware of who she was and how her very presence was no more than a confirmation of the aura's desparate need for something—anything—to keep it company.

—Brian Rotman, *River City* 20 (2000)

Introduction

LETTERED SELVES AND BEYOND

Reflecting on the relation between the human and the machinic, the cognitive theorist Andy Clark urges "We shall be cyborgs not in the merely superficial sense of combining flesh and wires, but in the more profound sense of being human-technology symbionts: thinking and reasoning systems whose minds and selves are spread across biological brain and non-biological circuitry." (2006, 1)

Until recently it was possible to believe otherwise. To believe that human organisms and their technologies, however messily intermixed and interdependent, were fundamentally different and in principle could be separated: on the one side, mind and culture and things of the spirit, on the other, tools, machines, and techno-apparatuses; the former invents and uses technology but is not itself, in its 'essence,' technological. Now, as technological systems penetrate every aspect of contemporary culture, bringing about an escalating and radical series of cognitive and social upheavals, it has become clear that no such separation of mind and machine is possible. Nor was it ever. Humans beings are "natural born cyborgs"; the 'human' has from the beginning of the species been a three-way hybrid, a bio-cultural-technological amalgam: the 'human mind'—its subjectivities, affects, agency, and forms of consciousness—having been put into form by a succession of physical and cognitive technologies at its disposal.

Leaving language aside for the moment, which properly speaking is a bio-cultural given rather than a technological medium, the chief mind-constituting technology, "mind upgrade" in Clark's phrase, and the mother of almost all subsequent cognitive upgrades, is writing. Writing in its two dimensions: the writing of ideas, patterns, and procedures whose most focused and abstract realization is the symbolic ecology of mathematical thought, and writing as an apparatus for inscribing human speech and thought among whose multiform achievements is the production of a literature and of a literate form of discourse, that enables one to read and write texts—such as the present essay—about the nature of writing.

Though I shall touch on some aspects of mathematical writing, my main concern will be with writing as the inscription of spoken language. For Western culture the writing of speech has been exclusively alphabetic, a system which from its inception has served as the West's dominant cognitive technology (along with mathematics) and the medium in which its legal, bureaucratic, historical, religious, artistic, and social business has been conducted. The result has been an alphabetic discourse, a shaping and textualization of thought and affect, a bringing forth of a system of metaphysics and religious belief, so pervasive and total as to be—from within that very discourse—almost invisible. Certainly, for at least the last half millennium the very concept of a person has adhered to that of a 'lettered self,' an individual psyche inextricable from the apparatus of alphabetic writing describing, articulating, communicating, presenting, and framing it. "In the society that has come into existence since the Middle Ages, one can always avoid picking up a pen, but one cannot avoid being described, identified, certified, and handled—like a text. Even in reaching out to become one's own 'self,' one reaches out for a text" (Illich and Sanders 1988, x). The "text," as Steven Shaviro observes, is the "postmodern equivalent of the soul" (1995, 128), a fact only recognized within the newfound interest in alphabeticism over the last two decades.

In the nineteenth century the totality of the alphabet's textual domination of Western culture encountered its first real resistance, its monopoly challenged by new media, technologies of reproduction and representation that have since appropriated many of the functions which had so long been discharged and organized by the alphabetic text: thus the alphabet's hold on factual description and memory was broken by photography; its inscription and preservation of speech sounds eclipsed by the direct reproduction of sound by the phonograph and its descendants; its domination of narrative form, fictional and otherwise, upstaged by documentary and film art; and its universal necessity weakened by television's ability to report or construe the social scene, via images and speech, in a manner accessible to the non-literate.

But this dethroning of the alphabetic text is now entering a new, more radical phase brought about by technologies of the virtual and networked media whose effects go beyond the mere appropriation and upstaging of alphabetic functionality. Not only does digital binary code extend the alphabetic principle to its abstract limit—an alphabet of two letters, o and 1, whose words spell out numbers—but the text itself has become an object manipulated within computational protocols foreign to it. At the same time the text's opposition to pictures—its ancient iconoclastic repudiation of the image—is being reconfigured by its confrontation with the digitally produced image. With the result that technologies of parallel computing and those of a pluri-dimensional visualization are inculcating modes of thought and self, and facilitating imaginings of agency, whose parallelisms are directly antagonistic to the intransigent monadism, linear coding, and intense seriality inseparable from alphabetic writing.

On a different (but ultimately related) track is the alphabet's reductive relation to the corporeal dimension of utterance, to speech's embodiment. Not only are letters in no way iconic, their visual form having no relation to that of the body or to how the sounds produced by the body's organs of speech are received by those hearing them, but the sounds which the letters notate are meaningless monads, minimal hearable fragments of speech absent any trace of the sense-making apparatus of the body producing them. This disconnect between alphabetic writing and the speaking body occurs most radically at the level of the phrase and the utterance. For what the alphabet eliminates is the body's inner and outer gestures which extend over speech segments beyond individual words. Both those visually observable movements that accompany and punctuate speech (which it was never its function to inscribe) and, more to the point, those inside speech, the gestures which constitute the voice itself—the tone, the rhythm, the variation of emphasis, the loudness, the changes of pitch, the mode of attack, discontinuities, repetitions, gaps and elisions, and the never absent play and musicality of utterance that makes human song possible. In short, the alphabet omits all the prosody of utterance and with it the multitude of bodily effects of force, significance, emotion, and affect that it conveys. It was the recognition of the reductive consequences of this omission, evident early in Greek literacy as soon as speeches were delivered by orators rather than the bards who composed them, that was instrumental in founding the art of Gorgian rhetoric. Since then confronting it has been the driving force in the historical development of all forms of 'prose' and poetic diction along with the reading and writing protocols associated with them.

It is not by chance that this previously ignored gestural dimension of speech should now be of interest. Among other things (its role in Sign language, the discovery of an intimate association between gesticulation and narrative speech, its relevance to voice-recognition software), gesture and gestural communication (to include haptic and tactile modalities) have become of growing importance within contemporary explorations of body/machine interfaces. A significant component of this reappraisal is the development of motion capture technology, a new digital medium which works by tracking the positions of markers attached to the moving body and recording their paths through three-dimensional space. What is captured as a digital file can include any kind of human (or animal) movement from dance, sport, and theatrical performance to the postures and passing gestures of social interaction. As such, the technology offers the possibility of capturing the entire communicational, instrumental, and affective traffic of the mobile body—projecting the outlines of a gesturo-haptic medium of vast potential. One whose theoretical significance has yet to be thought through, but whose practical realizations are already to be found in art objects and installations, animated film, computer gaming, electronic dance performance, and attempts to create virtual theatre. As alphabetic writing segmented the flow of speech into separate, decontextualized, discrete, and fixed objects of awareness—'words'—that could be examined and compared, giving rise to grammar, its own form of literate awareness, and the study of the resulting written language, so motion capture likewise opens the possibility of a 'gesturology,' a science of gesture that might allow the semiotic body the conceptual space to emerge from under the shadow of spoken language's lettered, disembodied inscription.

This is not of course to proclaim (which would be absurd) the imminent demise of alphabetic writing, or to want for communication in general the equivalent of what Artaud desired in particular for a theatre freed from subservience to written texts—the pre-eminence of screams, silences, and above all the gesturing body as the superior and proper vehicle for theatrical affect. But rather to point to the end of writing's three-millennia hegemony as the result of its ongoing subsumption within a digitally expanded mediational field. It is not, then, its still important and widespread use, but the *regime* of the alphabet that appears to be drawing to a close, the "period of alphabetic graphism" in André Leroi-Gourhan's phrase giving way to an era in which the inscribing of speech sounds with letters is but one element, not necessarily the overriding one, in the ongoing bio-cultural-technological 'writing' of the body's meanings, expressions, affects, and mobilities.

In the process, the alphabetic self, the embodied agency who writes and reads 'I,' and in so doing performs a complex play of same and otherness, actuality and virtuality, with the one who speaks and hears 'I,' will be confronted by a third 'I,' a self coming into being to the side of the written form, what might be termed a *para-self*, whose enunciation of 'I' will take place (and only take place if it is not to collapse back into its written predecessor) in the interior of a post-, better, trans-alphabetic ecology of ubiquitous and interactive, networked media.

"Writing," Walter Ong insists, "alters consciousness." (1982, chap. 4) Indeed. As do all media, not least each strand of the lattice of communicational technologies currently dissolving writing's pre-eminence, loosening the alphabet's hold by substituting *their* hold on consciousness. As Félix Guattari has observed, informational and communicational technologies "operate at the heart of human subjectivity . . . within its sensibility, affects and unconscious fantasms." (1995, 4) An observation that repeats the inescapable two-way intimacy remarked earlier between *techne* and *psyche*. Technology's mode of operation at "the heart" of the subject is not simply the action of something external introduced into a 'natural' psyche, one that was inner, private and secluded from technological influence. The operation of machines both augments already existing sites of technological mediation of the self and is transformative and works to constitute the very subject engaging with them. A phenomenon Roland Barthes observed for the action of writing: "In the modern verb of middle voice 'to write,' the subject is constituted as immediately contemporary with the writing, being effected and affected by it. . . ." (1986, 18)

This understanding of technology rejects the instrumental view of it as the use of tools and body-extending prostheses by pre-existing human subjects fully articulated before its deployment. And it likewise rejects the conception of technological media in terms of their representations, in terms of their content, the intentional manifest meanings they signify— whether linguistically, aurally, pictorially, kinetically, haptically—to pre-existing, self-sufficient subjects. In both views the phenomenon that is unseen and unexamined is the direct effect of technology's materiality, an effect always outside its explicit human, socio-cultural character and which transforms the bodies, nervous systems, and subjectivities of its users. This action of technology's 'radical material exteriority,' the subject-constituting work it performs, occurs at a pre-linguistic, pre-signifying and pre-theoretical level. As such it is antagonistic to understanding technology's achievement in terms of its purely discursive, socio-cultural constructions. Communicational media and semiotic apparatuses never coincide with their intended social uses or cultural purposes or their defined instrumentality or the effects sought and attributed to their manifest contents. Always something more is at work, a corporeal effect—a facilitation, an affordance, a restriction, a demand played out on the body—which derives from the uneliminable materiality and physicality of the mediological act itself, and which is necessarily invisible to the user engaged in the act of mediation.

Expressed differently, no encounter with a mediating apparatus can be reduced to the purely mental, ideational effects, one articulated within the discourse of its declared signifying and representational means, that occludes its physiological actions. As Steven Shaviro puts it for the apparatus of film, "We neglect the basic tactility and viscerality of cinematic experience when we describe material processes and effects, such as the persistence of vision, merely as mental illusions. Cinema produces real effects *in* the viewer, rather than merely presenting phantasmatic reflection *to* the viewer." (1995, 51) What is true of the "psychophysiology of cinematic experience" holds for any encounter with a mediating apparatus—cinematic, computational, telephonic, televisual, photographic, audiophonic, telegraphic, or any other: always the user is used, the psyche-body of the one who views, listens, speaks, computes is activated and transformed by an undeclared affect, a force outside the apparatus's explicit instrumentality.

Ignoring the action of this material underbelly projects an account of technological mediation that denies it an unconscious, denies it any under-the-radar, pre-discursive or pre-semiotic effects, and embeds its action and mode of being entirely in language and discourse, thereby domesticating it as a project and set of processes wholly capturable and able to be made explicit within conscious, representational thought. Mark Hansen describes this reduction as a fall into *technesis*, a "putting into discourse" of technology, a body-denying move that, he claims, underlies twentieth-century thinking about the nature of technology and material agency (2000, 2004).

Writing, like any medium, is a re-mediation; it engenders a clutch of interconnected discontinuities in the milieu of what preceded it: a disruption of the previous space-time consensus of its users and an altered relation between agency and embodiment giving rise to new forms of action, communication, and perception. It introduced a domain of virtual, seemingly 'unreal' objects, entities that are without context, endlessly repeatable, and free to be reproduced at any time, place, and cultural situation. For the medium of writing these virtual entities are texts. To engage with them writing posits, as does any medium, a virtual user, an abstract reading/writing agency who or which is as distinct from any particular, embodied, and situated user as an algebraic variable is from the individual numbers substitutable for it, an agency who/which accommodates all possible readers and writers of texts regardless of how and when in space and time they have or might have appeared. This floating entity makes ideas of disembodied agency, action at a distance, and thought transference plausible. As a result all communicational media have about them an aura of the uncanny and the supernatural, a ghost effect which clings to them. In the case of the telegraph, which introduced a new form of written converse with an absent agent, the effect conjured not ghosts as such; rather it inspired a new religion based on a telegraphic, table-tapping mode of conversation with the newly conceived ghost-spirits of the departed.

Long before this, writing (which had always been friendly to messages and self-proclamations from the dead) conjured into being ghosts of a different sort. Unlike telegraphy the conjuring did not follow immediately on the medium's deployment; it depended on a specific phenomenon—a self-reference within or by the medium, a written 'I'—to bring it about. Writing 'I,' pointing to a self in writing, is in effect making writing circle back onto the writer and confronting the self with a virtual simulacrum. Unlike the spoken 'I,' chained to its utterer with its referent unambiguously the one speaking, who or what the written 'I' is is indeterminate. It could be real or fictional, existent or nonexistent. It could be any writer of a text anywhere at any time for any purpose, a hypostatization or entification of the alphabet's virtual user: an unembodied being outside the confines of time and space operating as an invisible and unlocatable agency.

A trio of entities answering this description, namely God, Mind, Infinity, have formed the metaphysical horizons of Western religious, philosophical, and mathematical thought. Each such ghost is a phenomenon inseparable from alphabetic writing. The first arose from the writing of 'I' as in "I am the Lord thy God" and "I am that I am" to refer to and define Himself; the second from the writing of 'I' in Greek philosophical thought to refer to an un-embodied *psyche* lodged in the soma; the last from the writing of 'I' as a pronoun designating Aristotle's *nous*, that disembodied organ of rational thought able to count without end.

To summarize: A succession of media—speech, alphabetic writing, digital writing—each transforming their environments through a wave of virtuality specific to them. In the first, virtuality is located within the symbolic function per se, inherent in a speaker's capacity to refer to nonexistent and disembodied agencies; in the second, virtuality is located in writing's ability to signify across space and time in the absence of a real or embodied speaker; the third, still breaking, wave is manifest in the contemporary phenomenon of virtualizing X, where X ranges over the characteristic abstractions and processes of the alphabetic, pre-digital age. Associated with each of these virtual waves is a potential ghost effect, one specific to the medium concerned, realized in relation to a self-enunciation expressed within or by the medium. For language it is the ghostly presence of the other in the spoken 'I' giving rise to the belief in a 'spirit' separate from the gestural 'I' inseparable from the proprioceptive body; for alphabetic writing it is a transcendental agency, the hypostatizations we call God, Mind, and the Infinite Agent. For the digital or better network 'I,' a self-enunciation specific to contemporary media ecologies is still in flux, so a ghost effect, identifiable as a stable and repeatable phenomenon invoked by it cannot yet emerge.

However, within the contemporary digitally enabled scene, a network 'I' is being heralded. The features of such a third self-enunciating agency, differentiating it from the oral and scriptive 'I's, are becoming discernable. Such an 'I' is *immersive* and *gesturo-haptic*, understanding itself as meaningful from without, an embodied agent increasingly defined by the networks threading through it, and experiencing itself (notwithstanding the ubiquitous computer screen interface) as much through touch as vision,[1] through tactile, gestural, and haptic means as it navigates itself through informational space, traversing a "world of pervasive proximity" whose "dominant sense . . . is touch" (de Kerckhove 2006, 8). Such an 'I' is *porous*, spilling out of itself, traversed by other 'I's networked to it, permeated by the collectives of other selves and avatars via apparatuses (mobile phone or e-mail, ambient interactive devices, Web pages, apparatuses of surveillance, GPS systems) that form its techno-cultural environment and increasingly break down self–other boundaries thought previously to be uncrossable: what was private exfoliates (is blogged, Webcammed, posted) directly into the social at the same time as the social is introjected into the interior of the self, making it "harder and harder to say where the world stops and the person begins." (Clark 2006, 1) Lastly, such an 'I' is *plural* and *distributed*, as against the contained, centralized singularity of its lettered predecessor; it is internally heterogeneous and multiple, and, like the computational and imaging technologies mediating it, its behavior is governed by parallel protocols and rhythms—performing and forming itself through many actions and perceptions at once—as against doing or being one thing at a time on a sequential, predominantly endogenous, itinerary. In short, a self becoming beside itself, plural, trans-alphabetic, derived from and spread over multiple sites of agency, a self going parallel: a para-self.

Mental pathways, ways of believing, modes of thinking, habits of mind, an entire logic of representation, born from and maintained by alphabeticism over the last twenty-five hundred years, become increasingly incompatible with such a self. Metaphysical claims by religions of the book, authenticated by the assertions of an absolute monobeing from within an alphabetic text, become less tenable as their uncompromising insistence on an aboriginal singularity confronts the pluralizing, dispersive vectors of contemporary mediation. The West's ontotheological metaphysics, with its indivisible, unique-unto-themselves and beyond-which-nothing monads of an absolute Truth behind reality and a monolithic transcendent God entity begins to be revealed as a mediological achievement—magnificent but no longer appropriate—of a departing age.

PART I

One

THE ALPHABETIC BODY

The Alphabetic West

For Victor Hugo "Human society, the world, the whole of mankind is in the alphabet." (quoted in Ouaknin 1999, 9) Not quite. The Chinese system of writing speech is logographic: its characters notate morphemes, the smallest meaningful sounds, rather than the alphabet's meaningless phonemes. The Japanese use a mixture of morpheme- and phoneme-based systems. Neither of these cultures figured largely in Hugo's view of the world, but for Western civilization his trumpeting of the alphabet makes perfect sense: each of the two originating worlds, Judaic and Greek, which have respectively determined the West's religio-ethical and techno-rational/artistic horizons, was indeed created out of an encounter with a system of alphabetic writing.[1]

The encounters could not have occurred in more different social, historical, cultural, economic, religious, and intellectual milieus: 'cattle-herding semi-nomad' Israelites against slave-owning denizens of the Greek polis; agricultural exchange versus a monetized economy; scribe-priest control of writing versus a distributed citizen literacy; tribal kingdoms versus the militarized city-state; fixation on a single written corpus defining a religio-ethnic identity against an expanding ecology of literary and philosophical writings.

The Israelite encounter produced the transcendental Jewish God inhabiting a holy text, the sacred scroll of the Talmud or Five Books of Moses, a "library" of texts comprising "the verse of nomadic people, popular and religious songs of all sorts, mythical tales based on the cosmogony of the Middle East, oral traditions concerning national origins, prophecies, legislative and sacerdotal documents bearing . . . liturgical pieces, annals or chronicles, collections of proverbs written down long after their first appearance, . . . tales and romanticized fiction." (Martin 1994, 103–4)

The Greek encounter produced theatrical mimesis, deductive logic, and an invisible, disembodied Mind which has since its inception determined the relation of 'thinking' to 'writing' embedded in and transmitted by the founding texts of Western philosophical discourse. Each of these encounters and their metaphysical import will occupy us later (chapter 5).

Different alphabets were involved. Greek (its Romanized form now worldwide) was created circa 800 BCE when the Greeks modified the Phoenician consonantal alphabet by adding letters for vowels plus some consonants; Hebrew, used by the Israelites from circa 1000 BCE, thought also to be derived from Phoenician was, like it, voweless. Whereas vowels were necessary to inscribe Greek, a language which used them to register grammatical differences, Hebrew, a tri-consonantal semitic language, could be written without them. Plainly, the two alphabets will involve different writing and reading practices and be amenable to different uses.[2] Being entirely phonetic, the Greek alphabet allowed a word to be read outright from the text, whilst the Hebrew required interpretive work to determine it from the others within the semantic family indicated by its triple of consonants. For Ivan Illich and Barry Sanders, the former "picks the sound from the page and searches for the invisible ideas in the sounds the letters command him to make," and the latter "searches with his eyes for inaudible roots in order to flesh them out with his breath." (1988, 13) They suggest the Old Testament command by God to Ezekiel to breath life (or soul, *nefesh*) into the dry bones "so that they may live" is a metaphor for the need to add the moistness of vowels to lifeless consonants. More extravagantly, David Porush claims that all that is intellectually significant about the accomplishments of the Jews stems from this failure to notate vowels; an "imperfection" he connects to the "central metaphysical tenet" of Judaism, the "unpronounceability, the unwritability, and the unthinkability of the name of God." (1998, 54)

For this essay, the metaphysics of alphabetic writing, both Hebrew and Greek, will be seen from a perspective which doesn't turn on the presence or absence of vowels, or on the supposed travails of reading and interpreting an 'imperfect' script, or on the unpronounceability and so on of God's name (and its supposed metaphysical consequences), though all raise interesting issues, but rather on a feature of writing that precedes such phenomena, namely its ability, in its capacity as a medium, to perform a reflexive, self-citational move — inherent in the writing of 'I' — and thereby give rise, under appropriate conditions, to a disembodied, supernatural agency.

But before disembodied agencies come embodied ones. Alphabetic writing, like all technological systems and apparatuses, operates according to what might be called a corporeal axiomatic: it engages directly and inescapably with the bodies of its users. It makes demands and has corporeal effects. As a necessary condition for its operations it produces a certain body, in the present instance an 'alphabetic body' which has relations (of exclusion and co-presence) with existing semiotic body practices. The alphabet does this by imposing it own mediological needs on the body, from the evident perceptual and cognitive skills required to read and write to the invisible, neurological transformations which it induces in order to function. It is the latter effects, beneath the radar of the alphabet's explicit function of inscribing speech and so quite separate from its manifold inscriptional activities, that will be significant.

I shall approach the alphabetic body through the topic of gesture. The particular motive for proceeding thus will emerge in due course, but in relation to the general question of embodiment, communication, and human subjectivity the idea is not unnatural: there are deep-lying lines of force between gesture and becoming human. As an affective medium of the body and its semiotic envelope, gesture reaches deep into human sociality through its vital role in hominization (the proffered breast, the use of facial expressions, pointing, cuddling, the phenomenon of turn taking, the induction via visual capture and motherese into speech), and through its linkage to the embodied wordless empathy, the psychic mirroring of each other necessary for meaningful utterance and without which what sociologist Michel Maffesoli (1994) calls *puissance*, the 'will to live,' would not be possible. For Giorgio Agamben gesture constitutes a key category in relation to political ontology, a third term between means (pure action) and ends (pure production) whose essential mode of action is that within it something is "being endured and supported"—activities which, he claims, allow the "emergence of being-in-a-medium of human beings and thus opens up the ethical dimension for them." (2000, 57–58)

Human Gesture

Notwithstanding its role in empathy, hominization, and its relation to the ethical, making gesture the point of entry into the alphabetic body might seem puzzling. After all, the alphabet inscribes speech, and compared to the latter gesture is widely held to be crude and pantomimic, an atavistic, semantically impoverished mode of sense making overtaken by the development of language. And though evidently important in ceremonies and rituals, prayer, and sacrifices to gods, and crucial to all forms of dance, music, and theatrical performance, gesture would seem to offer little to any contemporary discourse on language, the nature of thought, and the technology of writing.

Such a diminished status is no longer the case.[3] Nor was it always so. In the middle of the seventeenth century John Bulwer, pursuing Francis Bacon's dream of discovering mankind's original language that disappeared in the Biblical catastrophe of Babel, turned to gestures, "transient hieroglyphs" Bacon had called them, as the key to the search. Bulwer, a physician, was interested in gesture's physiological character. He looked to the fact and manner of gesture's evident embodiment to provide clues to the original but now lost universal language. Bulwer, inventor of the first finger-spelling alphabet, opens his book *Chirologia* with an extraordinary tribute to the hands' abilities to convey meaning and incite affect:[4] "With these hands," he says, "we sue, entreat, beseech, solicit, call, allure, entice, dismiss, grant, deny, reprove, are suppliant, fear, threaten, abhor, repent, pray, instruct, witness, accuse, declare our silence . . . ," (1664, 8) and so on, for some two hundred manual signs—revealing a gestural microcosm of mid-seventeenth-century English social, religious, and legal encounters. In an earlier essay, *Panthomyotomia*, Bulwer attempts a metaphorical dissection of the muscles of the face and head in an attempt to relate their movements to the motions of thought taking place so near them. Bulwer's writings inaugurate a (yet to be consummated) gesturology and make him the first theoretician of the semiotic body. In the next century others followed, most famously Condillac with his attempt to lay out the gestural roots of language, Charles de Brosses's project on the gesturo-physiological origins of language, and the Abbé de l'Épée's championing of a language for the deaf composed of gestures.

But by the mid- to late-nineteenth century gesture had fallen victim to a scientific psychology which subordinated an emotionalized (implicitly feminine), gesturing body to a rational, speaking mind. A cruel consequence of this was the banning in 1880 at a conference of deaf educators in Milan of all use of Sign (gestural language) from European and American schools in favor of enforced voicing and lip reading by the deaf: "Gesture," the organizers insisted, "is not the true language of man. . . . Gesture, instead of addressing the mind, addresses the imagination and the senses. Thus for us, it is an absolute necessity to prohibit that language and to replace it with living speech, the only instrument of human thought." (quoted in Lane 1984, 391) Some eighty years later this phonocentric dismissal of Sign started to collapse when the gestural systems used throughout the world by the deaf to commune with each other were recognized as full-blown languages, on a grammatical, morphological, and semantic par with and in some respects superior to human speech. One consequence of this reevaluation of Sign was a reemergence of theories proclaiming the gestural affiliations and origin of human language.[5] (Nevertheless, some thirty years after Sign's linguistic recognition, traces of the phonocentric and textocentric derogation of gesture remain: several universities in the United States refuse Sign as a fulfillment of graduate language requirements on the kettle-logic grounds that American Sign Language [ASL] is not a 'real' language; ASL is not a 'foreign' language; and, in any case, ASL lacks a written form.) But its ability to replace the tongue as the vehicle and physical means of language is not the deepest nor, for our purposes, the most significant aspect of the relation between gesture and speech.

Interestingly, Maxine Sheets-Johnstone observes that the "skeptical assessments of sign languages, not to say their derision" are tied to the fact that in all forms of Sign the gestural articulations of thought are *perceived* rather than, as they are in verbal languages, *apperceived*. This fact makes mind–body dualists for whom thought is invisible and mental—inside the head—uncomfortable and reluctant to grant Sign the status of a language. (2002, 157) In relation to the body and alphabetic writing of spoken language gesture operates in the interior of speech itself as the presence of the body within utterance and the affective, intra-verbal dimension of the voice itself. But before this, a necessary clarification of the speech/gesture nexus by way of distinguishing two kinds of gesture: emblems and gesticulations—each with its own relation to language.

Emblem Gestures

Like spoken words, ASL gestures are coded entirely by a linguistic system. Distinct from these, not captured by a code, forming at most only a "partial code" situated between the two linguistic systems, is the field of so-called emblems. Emblems are what we ordinarily mean by 'gestures': holding up the palm, jerking the thumb, kissing one's fingertips, pointing, snorting, smacking one's forehead, squeezing a shoulder, bowing, slapping someone on the back, giving the shoulder, biting a knuckle, flourishing a fist, tapping the nose, shrugging, chuckling, beating one's breast, giving the finger, winking, and innumerable other visible, haptic, auditory, and tactile disciplined mobilities of the semiotic body.

According to studies initiated by David Efron (1941/1972) and Adam Kendon (1972), and subsequently developed by David McNeill and others, emblems are gestures whose principal function is to carry out certain social activities. "Emblems," McNeill writes, "are complete speech acts in themselves, but the speech acts they perform are restricted to a certain range of functions. They regulate and comment on the behavior of others, reveal one's own emotional states, make promises, swear oaths [and are] used to salute, command, request, reply to some challenge, insult, threaten, seek protection, express contempt or fear." (1992, 64) This list (that could easily be describing a portion of Bulwer's enumeration of the expressions of the hand) makes it clear that emblems are social, experiential, and interpersonal, deployed to make something happen, to impinge on the behavior of the self and others; emblems are not really interested in making statements, analyzing matters, or conveying facts and propositions.

Unlike speech they do not combine via a syntax as part of a language or an elaborated code. And they differ from words in that their meanings are neither explicitly defined nor (outside of instruction in rhetoric or acting) are they intentionally learned or studied, but rather they are picked up, absorbed and inculcated, taken in directly by the body, as it were, and (perhaps for this reason) remain stable in form and import over long periods of time despite linguistic changes in the communities of their users. These features indicate that emblem gestures might operate according to a different dynamic and logic, and might accomplish different ends, from those of speech. Calling them 'speech' acts, suggests they are within the horizons of speech and assumes they operate, as a mode of meaning or affect creation in the same ways and for the same purposes as speech. But is this so? Are emblems in any sense translatable into spoken language? Can they be transposed into words? What, for example, is the speech equivalent of a wink? Or, for that matter, a shrug? a slap on the back? folding one's arms? hands clasped in prayer? And do their mode of operation and outcomes resemble those of speech? If so, why as speaking beings do we bother with them?

The cultural range, robustness, and persistent use of emblems, their way of refusing and displacing speech, calls for an explanation. McNeill offers one in terms of 'word magic.' "Spoken words are special and carry with them the responsibility for being articulated. However, conveying the same meaning in gesture form avoids the articulatory act and, thanks to word magic, this lessened responsibility for speaking transfers to the speech act itself." (1992, 65) Doubtless, there is truth in the idea that gesturing rather than talking removes one from the net of justifications, arguments, questions, deceptions, interpretive qualifications, and recriminations that speech immediately introduces. But how many emblem gestures admit of the same meaning as a word or spoken phrase? Indeed, what does "having the 'same' meaning" mean? How convincingly can speech render an emblem? Giving the finger, for example, carries a different charge, has a different meaning, enables a different affect, initiates a different confrontation from *saying* "up yours" or "fuck you" or "go screw yourself," and so on. (That there are inequivalent verbalizations suggests emblems generate meanings by their very exclusion of speech.) But in any event, assuming that 'sameness' of meaning makes sense, is the difference between gesturing and voicing the 'same' meanings reducible to "lessened responsibility"?

Thus, consider other deployments of emblems, for example their extensive, deeply embedded, and seemingly indispensable use in secular and religious rituals and practices. Here something different from lessened responsibility, almost the opposite, seems to be in play; as if words, so easily uttered, are insufficiently responsible, not binding enough, too fleeting and precise at the same time, and only bodily action can fulfill the relevant ceremonial and devotional or liturgical purposes; as if gestures are able to create and stabilize belief, to induce as well as express religious feelings, social ideologies and moods, and forms of consciousness more radically and with more appropriate affect than the specialized precision of speech.[6] In this context, what André Leroi-Gourhan says about speech's (and writing's) inferiority to art vis-à-vis religion, "that graphic expression restores to language the dimension of the inexpressible — the possibility of multiplying the dimensions of a fact in instantly accessible visual symbols" (1993, 200), carries over from graphic symbols to visual gestures.[7]

Evidently, emblem gestures *say* nothing (even when they are auditory and even when they can be verbally parsed). In fact they function at their most characteristic when differentiated and opposed to speech. Unlike words, which stand in a coded relation to ideas, things, interpretants, people outside themselves, emblem gestures signify and have meaning — better: have force, affect, point — through the fact of their taking place, in the effects they help bring about, in the affectual matrices they support, in all that they induce by virtue of their *occurrence as events*. In other words,

emblem gestures do not say anything outside their own situated and embodied performance: their relation to speech is one of exclusion, avoidance, and on occasion silencing. Agamben locates the essence of gesture
in this silencing and articulates it as an exclusively metamedialogical phenomenon: "Because being-in-language is not something that could be said
in sentences, the gesture is essentially always a gesture of not being able to
figure something in language; it is always a *gag* in the proper meaning of
the term, indicating first of all something that could be put in your mouth
to hinder speech, as well as in the sense of the actor's improvisation meant
to compensate a loss of memory or an inability to speak." (2000, 59)

However illuminating it is to construe gesture in metacommunica-
tional terms, as the "making of a means visible," the formulation is ultimately reductive in several senses. First, as we've seen, emblem gestures
execute a variety of speech-act-like functions such as promising, threatening, and the like, as well as devotional and meditational acts that have little
or nothing to do with the "exhibition of mediality." Second, insofar as they
metacommunicate in this way, it is as emblems that they do so and not as
gestures at large; moreover, they behave in this way in relation to speech
and not necessarily with respect to other media. Third, even in relation
to speech, gesture behaves in ways other than a gag: besides excluding or
silencing speech or marking its inability to articulate in sentences the state
of being inside language, gesture co-originates with and accompanies spoken language as *gesticulation*, and on a deeper level is intrinsic to speech as
tone or *prosody*, the auditory gestures of the voice, without which human
verbal utterance is impossible. Lastly, characterizing gesture in exclusively
metalogic terms, "what is relayed to human beings in gestures is . . . the
communication of a communicability" (Agamben 2000, 58) masks the fact
that gesture is also and always a medium of no small importance in its own
right. To say more we need first to describe its gesticulatory and prosodic
forms of mediation.

Gesticulation

A casual look at conversation and storytelling shows verbal utterance accompanied by fleeting, often barely discernible, seemingly idiosyncratic
and indefinite gestures of the fingers, hands, arms, shoulders, and face.
Gestures that appear to be connected, although how is not clear, to the
substance of what is being narrated. These gesticulatory movements are
not planned or consciously produced; they are involuntary and sponta-

neous and are for the most part unnoticed and communicationally super-fluous. Certainly, blind people have no difficulty comprehending speech, and people converse easily on the telephone, listen to recorded messages, and fully understand speech on the radio, without registering too much disturbance at the absence of any accompanying gestures. More than other kinds of gesture, gesticulation seems an unnecessary addendum to utter-ance, an echo perhaps of a pre-intellectual, pre-verbal form of communi-cation, having little to do with the articulation or expression of thought in speech.

Empirical investigations of the gesticulatory activity accompanying verbal narration suggest otherwise. Far from being epiphenomenal, a surface effect unconnected to the expression of thought, gesticulation re-lates to the semantic, pragmatic, and discursive aspects of speech in non-trivial ways, embracing various kinds of gesture accomplishing distinct semiotic tasks. There are iconographic gestures, for example outlining a square shape depicting literally a 'window' or metaphorically indicating a 'window of opportunity'; kinetographic gestures, for example miming handwriting to indicate 'writing' or a 'text' or 'literacy'; or bringing the hands together accompanying the expression 'an agreement was reached'; or gestures to mark an abstraction introduced into the narrative, for ex-ample cupped hands (the container metaphor in Western culture) when narration jumps out of the story being told and refers to its genre. There are also 'beat' gestures, brief on/off movements marking the word they accompany as significant not for its semantic content but for its discur-sive or pragmatic role, for example, a hand flick when a new character or theme or metalingual gloss is introduced into a story. Beside these self-contained or holistic gestures there are also contrastive pairs. For example, a straight-line gesture indicating a direct source of information against a curved one indicating information that is mediated; or evolution in time by a series of looping gestures in contrast to a straight line for a succession of stages having no element of transformation. And there is a class of ges-tures that realizes experiential meanings, from self-pointing to specifying times and places, that correspond to deictic or indexical terms in speech such as 'I,' 'you,' 'here,' 'now' which make essential reference to the physical circumstances of the speaker.[8]

Plainly, gesticulation (notwithstanding its communicative redundancy in most practical contexts) is linked to the words it accompanies at non-trivial levels of speech. Why and how this has come about is not as yet understood. One might attempt an explanation along the lines that gesticulation translates a prior version of the sentence that is uttered, or that gesticulatory movements are created to illustrate, amplify, or gloss speech as the latter is produced. That such explanations (which give causal priority to speech) are not feasible follows from the way gestural form and the meaning of the utterance it accompanies are connected: there is a tight temporal binding, accurate to fractions of a second, discovered to operate between gesticulation and speech. Any gesture has a preparatory phase, a stroke phase in which the gesture proper occurs, and a withdrawal phase. In gesticulation, the preparation precedes the word(s) it relates to while the gesture itself coincides exactly at the height of its stroke phase with the word(s) in question, after which gesture and words disperse together; a simultaneous anticipation, coincidence, and falling away only possible if gesture and words are produced together, only if they issue from something preceding each of them; only if, McNeill argues, there were some earlier linkage, a common 'origin,' in some sense pre-verbal and pre-gestural, to them both.

For McNeill (1992, 2005) this gesture–word nexus consists of a dialectic of opposed modes of representation: gestural (imagistic, holistic, and synthetic) and verbal (linear, segmental, and analytic); the final utterance being the result of an interaction between a relatively free, privately formed, individual gestural impulse and the rule-based, public, socially constrained demands of a linguistic system. In other words, thinking, at least insofar as it eventuates in speech, has its beginning in visuo-kinetic images which then become gesticulated and verbalized to form an utterance. One might note in this connection the linguist Wallace Chafe's analysis of verbal utterance as composed of 'idea units' corresponding to single 'thoughts'; the duration of each unit being about two seconds, more or less the time for a complete gesture to take place. (Chafe 1985) It is as if spoken thought starts life as a yet-to-be-realized gesture, an idea we shall encounter later in a more developed form operating within mathematical thought. I turn now to the third form of gestural mediation and speech, that which inhabits speech itself, namely prosody.

Gestures of the Voice

Emblem gestures operate outside of and alternative to speech; gesticulation operates alongside and parallel to speech. We come now to another form, ultimately more significant for our purposes, the audible body movements which operate *inside* speech: gestures which constitute the voice itself.

Speech involves systematic and interconnected movements of the lips, tongue, cheeks, jaw, glottis, vocal chords, larynx, diaphragm—identifiable and repeatable patterns of body parts—suggesting that it might be usefully regarded as a species of gesture, auditory as distinct from visual, but gestural nonetheless. Findings from research in phonetics and artificial speech synthesis over the past two decades confirm this. They indicate that it is precisely as a gestural system that the complex kinematics—the aural/oral assemblage of movements that make up the human voice—are best comprehended. Specifically, the task-dynamical, physiological models of the type describing the assemblage of movements and skeleto-muscular organization of the body during walking have proved ideal for modeling the dynamics of the lips, tongue, larynx, and so on, during speech production.

Moreover, not only is the production of speech gestural, but so it turns out, somewhat unexpectedly, is its perception: "Surprisingly," as the evolutionary neurologist Terrence Deacon finds himself saying, "auditory processing of speech sounds does not appear to be based on extracting basic acoustic parameters of the signal, as a scientist might design a computer to do, before mapping them onto speech sounds. Speech analysis appears designed instead to predict which oral-vocal movements produced them and ignore the rest." (1997, 14) We listen, it seems, not to speech sounds as such, not, that is, as isolatable sonic entities, but to the movements of the body causing them; we focus on what happens between the sounds, to the dynamics of their preparatory phases, pauses, holds, accelerations, fallings away, and completions—the very features of gestures we attend when we are perceiving them. In a certain sense, we listen to speech-sounds as signs of their gestural origins, as a physician listens to the sounds a patient's heart makes in order to analyze the movements causing them.

Linguists draw a fundamental division between two aspects of spoken utterance: they separate what is considered by them to be 'proper' to language—what is actually said, the grammatically and syntactically governed strings of phonemes, words, phrases, and so on, for example, "To be or not to be"—from the 'paralinguistic' manner of their saying, from how what is said is said, for example in this case, the prosodically varied ways such a line might be delivered by an actor playing Hamlet. Prosody is the gestural dimension of the voice, its "grain" (Barthes): it comprises all the vocal dynamics often referred to simply as 'tone,' or 'tone of voice,' namely the phrasing, the intonation, the musicality, the rhythm, the volume and emphasis, the rise and fall of pitch, the fallings away and accelerations, the pauses, gaps, hesitations, the anticipations, elisions, silences, elongations, repetitions, and contractions that the word-strings of an utterance are subject to.

Prosody has an ancient lineage. It originates from innate primate calls. But though the two are closely related as signals, and though "laughing, sobbing, screaming with fright, crying with pain, groaning, and sighing" constitute a more or less innate repertoire of prosody-like calls, the two are distinct.[9] For, "unlike calls of other species," Deacon points out, "prosodic vocal modification is continuous and highly correlated with the speech process. It is as though the call circuits are being continuously stimulated by vocal output systems" (1997, 418), as though, as a neurological consequence of hearing oneself speak, the midbrain and limbic systems responsible for primate calls become detached from instinctual control, become de-innate. This allows them to be eventually re-deployed in expressively and semiotically variable ways: the vocal gestures that constitute prosody become culturally malleable vehicles of human affect.[10]

According to a recent, neurological account of the evolution of language by Terrence Deacon (1997) (see chapter 5), this move, the escape from instinctual calls to what we now identify as the prosodic dimension of spoken utterance, did not occur overnight. The prosodic system which is essentially a "system of indices" that direct attention to what the speaker deems to be most salient, must have been "tightly linked to the evolution of speaking abilities," that is, to the trans-indexical, symbolic use of words, over a considerable time period, since the two systems are "parallel and complementary to one another anatomically as well as functionally." This deeply laid down parallelism, manifest as a "seamless complementarity" (1997, 364), rests on a neurological division of labor between control of the rapid-fire articulatory phonemic movements and the slower waves of prosodic gestures.

One might observe here a similarity to the paralinguistic activity of gesticulation. The split-second coincidence of words and their accompanying gesticulation—a consequence of their co-origination—is here a literal fusing: word and gesture are integral, two sides of the same utterance heard as a single acoustic event. However, in the case of gesticulation, the gestures signal meta-linguistic and discursive features of the ongoing verbal narration, which are essentially markers of cortical origin, whereas here the gestures constituting tone of voice signal subcortical, affectual aspects of the utterance originating in the midbrain, aspects that are often vital to the meanings put into play. As rhetoricians and actors know, differences in tone can make the same words gentle or withering, questioning or threatening, flattering, indifferent, or menacing, or sardonic or gleeful or seductive or pleading, and so on.

But, as we shall see, the inseparability of words and their tone, the seamless whole that constitutes verbal utterance, applies only to speech. It is precisely what is lost when writing enters the scene; as soon, that is, as utterance is transcribed and rendered as an alphabetic text. (Indeed it is only in the wake of writing that these separate aspects of speech appear.)

Writing Speech

What if one could separate speech from the voice? Eliminate the tone and keep the words? Alphabetic writing is a communicational medium, and every medium disrupts what had been for its predecessor conceived as a seamless whole, an integrated assemblage. The process of remediation involves a recalibration of space-time with consequent separations and severings of what were spatial and temporal and physical and aural contiguities and a reconstitution of (a dimension of) the original content in virtual form, which for writing is the text, speech being reconstituted in virtual form as 'speech at a distance.' Writing segments the spoken stream of sounds into words (which themselves owe their status as separate items to the action of writing) from the time, place, circumstances, psychological wherewithal, and social contexts of its production and re-situates it at another time, elsewhere, for other purposes, in other circumstances, in unknown contexts. It cuts speech loose from the voice, substituting for the individual, breathing, here-and-now agency of the one who utters them by an abstract, invisible author, and replacing a unique event, the utterance which unfolds over time, by fixed, repeatable, atemporal alphabetic inscriptions, inscriptions which necessarily fall short of representation. "Speech," Barry Powell observes, "is a wave," and the alphabet's separate graphic marks "cannot represent it." (2002, 123) And, more salient here, alphabetic writing eliminates all and any connection speech has to the body's gestures. One might object that handwritten alphabetic texts evade this total disjunction from gesture. Written emphasis, uncertainty, rhythm, discontinuity, stress, tailing off, and other scriptive traces of the body, might be said to be the handwriting correlates to certain rudimentary forms of vocal gestures. But the effect, to the extent it exists, is tenuous and not uniform enough to serve any reliable communicative function. In any event, it was effectively eliminated from public texts with the arrival of printing and increasingly from private ones by typewriting.[11]

At first glance this elimination of gesture is what one would expect. After all, the alphabet writes speech and has no truck with emblems, which operate outside the domain of speech, and it has no interest in gesticulation, which adds nothing in practice to speech and is thus irrelevant to the alphabet's task. But omitting the gestures that are interior to speech—eliminating the entire prosodic landscape of vocal gesture—is another matter, one which makes clear that the alphabet does not and in fact cannot write speech. Alphabetic letters don't capture or represent or notate the utterance that comes out of the mouth and is heard by a listener. They notate individual(ized) words, which (in the wake of writing) can, as we've seen, be designated simply as 'what's said,' but they do not notate the prosodic dimension, not the affect, force, point, and manner of delivery of the words, not how what's said is said.

It would be difficult to exaggerate the consequences of prosody's omission for the development of Western literacy: responding to it has been the condition for the possibility of this literacy, shaping and inventing what counts as a text and, what is the same thing, establishing the protocols of reading alphabetic writing. The recognition that such writing falls short of speech was of course immediately apparent early on: in the writing down of Greek funeral orations and the problem of their delivery by an orator other than their bardic author; and in the Jewish Talmudic tradition of endless rabbinical interpretation engendered by the problem of interpreting the 'spoken' word of God. A contemporary account of writing's inability to render tone is offered by speech-act theory: "Writing," David Olson writes, "lacks devices for representing the illocutionary force of an utterance, that is, indications of the speaker's attitude to what is said which the reader may use to determine how the author intended the text to be taken. The history of reading is largely the history of attempting to cope with what writing does not represent." (1994, 145)

This is at best a partial truth. More is involved in prosody's absence than the loss of illocutionary force, understood here in terms of a speaker's "attitudes" and author's "intention" as a deliberately formulated, linguistically explicit, consciously presented 'thought'; an abstract entity whose very conception within the written history of the West is of an item originating in a 'mind'; itself an abstract disembodied entity brought into being by alphabetic writing. Equating what writing omits with the content of a conscious speech act only obscures the inexplicit, a-conscious effect of this loss which, as we shall see, is key to the very construction of 'mind.' For the present we can observe that the identification is perforce reductive: it occludes the corporeal underside of the alphabet's action. What writing omits from speech is the body: the feelings, moods, emotions, attitudes, intuitions, embodied demands, declarations, expressions, and desires located in the voice, rather than consciously formulated (writable) thought. What it omits is the entire field of affect conveyed and induced by human vocality, through the voice's impulsions, inflections, and rhythms, its aural texture and emotional dynamics. A vocal field bordered on one side by song and on the other by the non-speech of sighs, moans, cries, grunts, screams, laughs, and so on, all that, in Roland Barthes's phrase, surrounds a "language lined with flesh." (1975, 66–67)

Notwithstanding this, it is still true that the history of reading is the history of redressing what writing fails to represent. Or, the same thing, the history of writing consists principally of attempts to find readable equivalents and alternatives to the vocal prosody necessarily absent from it. Lacking vocal gesture, writing was obliged to construct its own modes of force, its own purely textual sources of affect, which it accomplished through two dialectically opposed—or better, co-evolutionary—principles of creation: transduction (the discourses of narrative prose) and mimesis (the voices of poetic diction).

The poetic generation of affect is the more direct, iconic, and corporeally rooted one of mimetic retrieval: it seeks to recuperate vocalic gestures, to reproduce the oral/aural achievements of an embodied voice within the sound effects of speakable texts. Paul Zumthor commenting precisely on the performance and reception of oral poetry talks of "body movements" being "integrated into a poetics," and he notes the "astonishing permanence that associates gesture and utterance," insisting that "a gestural model is part of the 'competence' of the interpreter and is projected into performance." (1990, 153) The impulse is to reproduce a kind of sonically faithful simulacrum of the work of the voice: through words chosen as much for their aural/oral features as their significance, through fusions and splittings of phrases, through the deployment of textual arrangements—and ordering, juxtaposing, spacing, enjambment—which mimic the gesture-based dynamics of toned speech.[12]

With prose, retrieval gives way to textual reinvention. Here, alphabetic writing brought into being an entire apparatus of its own for inscribing affect. Prose rejects any directly sonic recuperation of vocalic gestures in favor of a textual transduction of them. It transposes or transmutes prosodic effects into inscriptional ones through the invention of new, textual forms governed by grammar and syntax rather than sonic values, in the process distributing (written versions of) affect across the entire lexical and syntactic landscape via the creation of a range of devices—neologisms, phrasal conventions, textual diagrams, rhetorical inversions, figures of 'speech,' letter-forms, and narrative formulas and 'styles.' These devices serve and facilitate a great variety of affectual desiderata for various purposes, from the literary project of inducing polyvalence and estrangement (the primary function of literature according to Victor Schlovsky) to the sought-after clarity and unambiguous neutrality of legal texts and the 'plain style' of scientific prose intended to eliminate any trace of 'subjective' and nonliteral affect.

Observe that, strictly speaking, the development of prose and poetic diction is not the fruit of the alphabet alone, in the sense of being constructed from letters. Both mimesis and transduction called for and in turn were forwarded by devices and techniques of punctuation that discharge a core set of functions handled orally by tone. These are extra-alphabetic having to do with handling text—blank spaces between words, commas, question marks, periods, quotation marks, paragraphs, hyphens, marks of ellipsis, capital letters, exclamation marks, parentheses—rather than representing sound elements of speech. Moreover their introduction went hand in hand with the conceptual innovations they offered, "Certain constructs that cannot exist without reference to the alphabet—thought and language, lie and memory, translation, and particularly the self—developed parallel to these writing techniques." (Illich and Sanders 1988, x) To this must be added the use of these devices to aid reading. Thus St. Jerome described the segmenting of texts, writing 'by clauses and phrases' (*per coma et commata*), that he had found in classical texts, as being more intelligible to the reader than the textual practices of his day: "It told the reader either to raise or lower the voice, in order to render sense through proper intonation." (Fischer 2003, 48) But, as we have seen, only a small portion of (the work of) intonation, proper or otherwise, has been built into the augmented devices of the alphabet. Interestingly, the process continues: a whole new generation of punctuation techniques—mark-up languages, scripting codes, and style sheets—specifically for augmenting electronic texts are now being developed and used to make them easier to be read aloud by voice-synthesizing machines.

The two sides of human speech, syllables (self-contained, discrete) and the tone (continuous and extended) of their delivery, are governed neurologically by the comparatively recent neocortex and the ancient midbrain or limbic area respectively. In speech—indeed as speech—they occur simultaneously, are united, coupled into a single meaningful sonic event. Alphabetic writing disrupts this unity. It splits the voice, selecting from the stream of speech (what it defines as) words to notate, jettisons all trace of their tone, and sets up its own neurological apparatus to handle the writing and reading of the resulting letter notations.

At this point a neurological qualification is in order. Here and in what follows I deliberately adopt a simplified topographic picture of the arrangements of the brain's functions. In reality there seems to be not one localized area of affect and feeling—the midbrain and limbic structures—and one localized area of cognitive thought—the neocortex—but a number of separate, interconnected regions, distributed throughout the brain, each specific to types of emotion and cognitive processing. This is part of an emerging consensus of the brain as "a collection of systems, sometimes called modules, each with different functions." (LeDoux 1996, 105) An adequate neurological picture of how affect and abstract thought, tone and 'what's said' are interrelated, then, would have to incorporate how the relevant modules are connected to and interact with each other. No such account at present exists. In fact neurological interest in the topic of affect is relatively recent and has yet to provide such a picture. However, for the purpose of identifying the rudimentary, but highly significant neurological effects brought about by the practice of alphabetic writing and reading, a crude first approximation, framed as an opposition of cortical and midbrain systems, is more than adequate.[13]

'Learning one's alphabet,' acquiring the ability to read and write alphabetic inscriptions, is an intense cognitive business requiring a permanent alteration of their brains that takes human children a protracted period of repetition and practice to accomplish. Neurologically, the requirements of literacy create in the brain what we might call a 'literacy module,' a neural complex within the neocortex dedicated to writing and reading purely textual entities, that is, handling the production and reception of phonemic strings that constitute written words shorn of their prosodic content and associated affective fields, words decoupled from the moods, feelings, desires, and regulatory activity routinely evinced (and induced) by spoken utterance. The module comprises a mesh of pathways centered in the frontal-occipital lobes and virtually unmoored from the midbrain. As such it is distinct from the 'speech areas,' the lateral-parietal network governing the generation and reception of utterance and which from the advent of language has been coupled to the affect apparatus of the limbic systems and midbrain.

One can relate the picture here to the neurological theory of 'emotional conditioning' put forward by the neurologist J. LeDoux (1994, 1996), according to which an input train of stimuli is split in two pathways: one going to the limbic systems and the other to the neocortex. The result is a division of labor: the older and earlier limbic systems govern the rapid affective evaluation of the stimuli with respect to memory and conditioning; the more recent neocortical apparatus handles the slower, context-dependent rational analysis of the stimuli. Though autonomous, the two pathways intertwine and combine affect and analysis within an emotional response. In like manner, the signifying neocortical dimension of words and their affective, prosodic dimension mediated by the limbic systems combine in speech. But writing's elimination of vocal affect foregrounds the neocortical dimension, which is thus set in opposition to the speech it purports to represent.

The opposition between speech and disembodied writing is a hierarchy. This is in the obvious sense that in the process of establishing itself as the vehicle for the creation and furtherance of Western culture, writing has from its inception dominated speech, assigning it a subordinate status within literacy's increasing colonization of all that was the province of oralism. And also in a less evident sense of being patterned on a neurological precedent difficult to avoid. Thus Terrence Deacon observes the production of human speech might be modeled on the "superposition of intentional cortical motor behavior over autonomous sub-cortical vocal behavior" necessary to counter the "unintended eruption of primate cries." (1997, 244) In like manner, writing can be seen as demanding a neocortical override of the midbrain, a superposition necessary to suppress or inhibit the production of prosodic speech. Corresponding to the unintended eruption of primate cries, then, one has the counterproductive eruption of vocal affect, of prosodic gestures which, interestingly in this connection, themselves derive from de-instinctualized primate cries. In any event, it seems that the familiar hierarchies—cognition over affect, thought over feeling, signification over force, and ultimately mind, soul, and spirit over body and soma—that permeate the intellectual mainstream and values of Western culture, might have their antecedents in an absent—better, disenfranchised and repressed—midbrain set against a consciously present, inevitably foregrounded and dominating neocortex.

This means that on the one hand, writing's de-prosodized words appear incorporeal, as if they issued from a disembodied and autonomous source.[14] On the other hand, from its beginning, writing has effaced its own role in constructing the hierarchies of mind over body, thought over feeling, and so on. By claiming (in writing) to re-present speech without loss, by systematically identifying itself as a medium which transparently inscribes speech, it masks the radical disjunction from speech that enables it to make such a claim.

Conflating virtual and actual speech has consequences of an ontological and metaphysical kind. Once the alphabetic body is in place, once the neuronal pathways of literacy have been installed in the brains of its users and became automatic through the repeated alphabetic writing of speech and reading of lettered texts, that is, as soon as writing "invisiblizes" itself as a medium, the stage is set for the coming into being of an entity—necessarily incorporeal—who is imagined to write 'I.' Such a being or agency 'speaks' itself with a virtual voice and, in (undeclared and unexamined) analogy to the spoken 'I,' is imputed to be the source and origin of virtual speech. In chapter 5, we shall see how two such agencies—Mind and God—exhibiting different modes of transcendental escape from corporeality can be understood as medialogically engendered ghosts, spectral quasi-presences which emerged out of the alphabetic writing of 'I.'

One can ask about the 'speech' of such virtual beings. For example, seeing that the connections essential to vocalic affect are routed through the frontal lobes, imagine the suppression of them as performing a kind of orthographic version of a pre-frontal lobotomy: certainly, descriptions of lobotomized speech, "in their words . . . no traces of affection could be detected" (Amaral and Oliviero 2005), suggest how, if it were possible to realize it, we might perceive de-prosodized words, speech emptied of all affect; an idea I return to later.

Two

GESTURE AND NON-ALPHABETIC WRITING

Gesture inside Mathematics

> Thinking is not a process that takes place 'behind' or 'underneath' bodily activity, but is the bodily activity itself.
> —Ricardo Nemirovsky and Francesca Ferrara, *New Avenues for the Micro-analysis of Mathematics Learning: Connecting Talk, Gesture, and Eye Motion* (2004)

Gesture's relation to alphabetic writing lies in the omission of prosodic affect and its subsequent textual retrieval and re-invention in poetry and prose. We come now to the action of gesture in relation to a species of *non-alphabetic* writing—the ideograms and diagrams which encode meaning in mathematics.

The last decade or so has seen an increasing focus on the importance of physical activity and bodily mechanisms within all forms of learning and thought, not least within the highly theoretical, abstract, and long-considered disembodied concepts of mathematics. Thus, when George Lakoff and Rafael Núñez contend that "Mathematics is embodied, it is grounded in bodily experience in the world" (2000, 365), or Núñez refers to "the embodied cognitive foundations of mathematics" (2006, 160) they are claiming that mathematics rests on a network of inferences derived from metaphors of basic body activities—such as starting, stopping, finalizing an action, continuing along a path of motion, gathering together a plurality of objects, etc.—which they understand as cognitive mappings and schemas underlying the various operations of elementary arithmetic, the theory of sets, and other mathematical abstractions out of which the bulk of mathematical concepts are formed.

Other, more directly empirical, studies take a more immediate route to the body's engagement with mathematical ideas by observing it in situ, as part of a discourse or performance, observing how, for example, the body behaves in the process of grappling with a new mathematical concept. Here, the question is not one of representation—the apparatus of embodied metaphors, similes, or metonyms supposedly 'behind' the mathematics—but on what is revealed by the physical activities themselves, the moving around, visualizing, talking, scribbling, and gesturing involved in learning and communicating the subject. Thus, by tracking the moment-to-moment eye movements (saccade gestures) of a group of mathematics students arguing about, notating, and engaging with real and imagined diagrams, Nemirovsky and Ferrara found the students' thinking to encompass "parallel streams of bodily activity" manifest as a "coordinated activity among hands, eyes, and talk in the process of expanding, or bringing into the open, aspects of visual meaning" (2004), an organic notion that leads them to concur with the thesis that "Children's thinking," and hence human thinking in general, "is more akin to an ecology of ideas, co-existing and competing with each other for use, than like monolithic changes from one stage of understanding to the next." (1–2)[1]

One consequence of this embrace of an ecologically understood thinking body is to establish that the deep links gesture has to speech and thought are not confined to verbally expressed narration, but appear to be significantly linked to the nonverbal ideograms and diagrams that comprise mathematical languages.

The idea that gestures of the body, abstract thought, and mathematical diagrams are intertwined is not new. The philosopher Maurice Merleau-Ponty, whose entire phenomenological project could be summarized as a meditation on 'the flesh that thinks,' was advocating a version of it half a century ago. Thus, starting with the presupposition that the geometer is dynamic and embodied, "The subject of geometry is a motor subject" (1962, 443), and confronted with the standard definition of a triangle as a three-sided figure, he insists that "There is no definition of a triangle which includes in advance the properties subsequently to be demonstrated," no "logical definition of the triangle could equal in fecundity the vision of the figure." (441) On the contrary, the creative force, their ability to mediate new meanings, of mathematical entities such as triangles is pre-formal, inseparable from our lived, embodied, and dynamic interaction with them. Before all else triangles are 'thought' through the active body. A triangle's essence is physical, concrete, a "certain modality of my hold on the world."

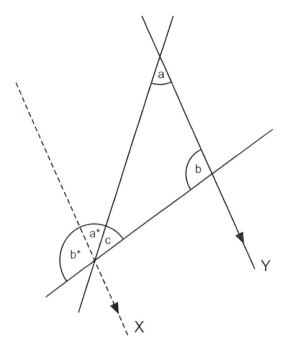

Figure 1. Merleau-Ponty's triangle.

(442) And this is literally so: a 'hold' consists for him of the drawing and perceiving gestures which determine any diagram, but which is never exhausted by them, since, as Merleau-Ponty points out, the gestures overflow any particular diagrammatic representation.

The standard geometrical proof that the angles of a triangle sum to 180 degrees goes as follows: a = a* and b = b*, since lines X and Y are parallel and any transversal intersecting parallel line does so at equal angles. From this we can conclude that a + b + c = a* + b* + c. But the right-hand side of this equation is the angle subtended by a straight line, which by definition is 180 degrees.

After presenting this demonstration Merleau-Ponty observes, "My perception of the triangle was not . . . fixed and dead . . . it was traversed by lines of force, and everywhere in it new directions, not traced out yet, came to light. In so far as the triangle was implicated in my hold on the world, it was bursting with indefinite possibilities of which the construction actually drawn was merely one." (443)

More recently, the mathematician Gilles Châtelet elaborated a mathematically far-reaching and more-sophisticated version of Merleau-Ponty's account of diagrams in his work on the mathematization of space (2000). For Châtelet, the fecundity of diagrams, their ability to mobilize what Merleau-Ponty sees as "indefinite possibilities" of mathematical meaning, derives precisely from their relation to gestures.

Two principles organize Châtelet's genealogy of physico-mathematical space. One is an insistence on intuition and premonition, on the a-logical, the 'metaphysical' or contemplative dimension of mathematical thought; the other is an insistence that mathematical abstraction cannot be divorced from "sensible matter," from the movement, perceptual activity, and agency of bodies. Their combination—an embodied rumination—is for Châtelet the generative source of mathematical ideas, a claim he demonstrates by examining the role that gesture and diagrams (they are for him inextricable) play throughout mathematics, and especially their function within the geometrization of space.

Diagrams, figures of contemplation, live for Châtelet in the aftermath of the body's gestural mobility: gesture at both an organizational and methodological level. Thus, speaking of the large-scale periodization of the historical dynamic forming the mathematics/space nexus, he declares: "Gesture and problems mark an epoch." (3) And in terms of method: "The concept of *gesture* seems to us crucial in our approach to the amplifying abstraction of mathematics": a mode of abstraction that cannot be paraphrased nor metaphorized nor captured by formal systems that "would like to buckle shut a grammar of gestures." The gesture is not referential, "it doesn't throw out bridges between us and things," and it is not predetermined, "no algorithm controls its staging." It would be better, he says, "to speak of a propulsion, which gathers itself up again as an impulse, of a single gesture that strips a structure bare and awakens in us other gestures." (9) He emphasizes that gestures refer to a "disciplined distribution of mobility before any transfer takes place: one is infused with the gesture before knowing it." (10) In other words, gesture is outside the domain of the sign insofar as signs are coded and call for a hermeneutics, an interpretive apparatus separable from, and in place prior to, the act of signification. Rather, the mode of action of gesture is enactive, exterior to anything prior to its own performance: it works through bodily executed events, creating meaning and mathematical significance "before one knows it."

One consequence of the foundational rigor imposed on the presentation of mathematics in the twentieth century has been an obscuring of its connection to the activity—gestural, visual, meditational—of the body. Characterizing mathematics as an edifice of formal, set-theoretical structures ignores the corporeality, the physical materiality (semiotic and performative), as well as the contemplative/intuitive poles of mathematical activity; and in so doing dismisses diagrams as mere psychological props, providing perceptual help but contributing nothing of substance to mathematical content.[2] Against this, Châtelet explains their ubiquity within mathematical practice by assigning a fundamental importance to the work they do in the creation of 'content': diagrams are more than depictions or images: they are frozen gestures, they "capture gestures mid-flight" and thus have a kinematic as well as a purely visual dimension. "A diagram can transfix a gesture, bring it to rest, long before it curls up into a sign," which is why, he says, "modern geometers and cosmologists like diagrams with their peremptory power of evocation."(10)

A crucial attribute of diagrams in Châtelet's understanding is their after-life, their capacity to be re-activated, to not get used up or "exhausted." This separates them from metaphors: they can "prolong themselves into an operation which keeps them from being worn out." True, like metaphors, they can "leap out in order to create spaces and reduce gaps," but unlike a metaphor, a diagram initiates another phase: when a diagram "immobilizes a gesture in order to set down an operation, it does so by sketching a gesture that then cuts out another." Châtelet thus understands the work of diagrams as a relaying of gestures, as a "technique of allusions," and speaks of dynasties of them, "families of diagrams of increasingly precise and ambitious allusions." (10) which extend through the historical development of mathematics; the recuperation of their lineage is one of the principal tools Châtelet uses to organize his genealogies.[3] Diagrams are not representations of existing knowledge or already available content; they distill action and experience and "reveal themselves capable of appropriating and conveying 'all this talking with the hands' . . . of which physicists are so proud." Châtelet sees the appeal to and deployment of diagrams as a perennial element of mathematical practice: "The diagram never goes out of fashion: it is a project that aims to apply exclusively to what it sketches; this demand for autonomy makes it the natural accomplice of thought experiments."[4] (11)

Mathematics, then, offers two modes of converting the "disciplined mobility of the body" into signs: transducing it via metaphors which "shed their skin" to become symbolic operations such as adding numbers; and capturing it, freezing mobility in mid-flight to form a diagram. (Recall our earlier encounter with these two modes as the means of engendering literate versions of vocal affect within alphabetic writing: prose, Châtelet's metaphor (transduced gesture), and the poetic, his diagrams (captured gesture). In each case the source of the mathematics is not itself mathematical; it arises from what Châtelet calls 'ruminative' or 'contemplative' and on occasions 'metaphysical' thought. It corresponds to what most mathematicians refer to simply (and opaquely) as 'intuition': the hunches, instincts, premonitions, convictions of certainty without evidence, and numerous other gut feelings that seem to hover over any engagement with the subject.

Attempts to account for these a-logical, pre-verbal feelings in terms of an epistemology of 'objective' mathematical objects usually end up in a form, more or less mystical, of Platonic metaphysics. Thus, for example, the mathematician Alain Connes proclaiming his belief in a mathematical realm that exists "independently of the human mind" insists that mathematics is the exploration of an "archaic reality"; the mathematician "develops a special sense . . . irreducible to sight, hearing or touch, that enables him to perceive a reality every bit as constraining as physical reality, but one that is far more stable . . . for not being located in space-time" (1995, 28). But talk of an 'archaic reality' is romantic and mystifying. If one accepts the embodied—metaphorical and gestural—origins of mathematical thought, then mathematical intuition becomes explicable in principle as the unarticulated apprehension of precisely this embodiment. And there is no reason why mathematicians' perceptions of their inner kinesis, autohapticity, and proprioception, should lie outside the space-time in which their bodies consciously or otherwise situate themselves. After all, such apprehensions are by no means strange: humans 'know' things; they embody knowledge of their environment that has enabled them to survive, gesture to each other, make tools, hunt other bodies (human and not), and roam the world long before the advent of speech and the configuration of consciousness that came with it. So that Connes's archaic reality, his intuition of it as the domain of mathematical objects, might be no more and no less than a felt connection to his body, the apprehension of his inner gestural activity, a perception, perhaps, of his own neurophysiology which, it seems, Connes (and others) is able to harness as a source of mathematical abstraction.[5]

The hand in the age of the image seems amputated, no longer able to cover the eyes or the face, overwhelmed by the landscape of speed. Yet, this frenzy of the surface provides the hand its cachet; touch has become synonymous with the genuine, the real, the human. Touch, is nostalgic. Touch cannot be mediatized, technologized, mechanized. Yet this dichotomy is false—or at least simplistic.

—Christof Migone, "The Prestidigitator: A Manual" (2004)

It is through the skin that metaphysics must be made to re-enter our minds.

—Antonin Artaud, *The Theatre and Its Double* (1958), 99

the curiously archaic present

Imagining a future in which alphabetic writing and with it philosophical essays and literature as we know them will disappear, replaced by forms evolved from them, André Leroi-Gourhan assures us at the end of *Gesture and Speech* that the mentality and accomplishments of these artifacts will not be lost, since the "curiously archaic forms employed by thinking human beings during the period of alphabetic graphism will be preserved in print." (1993, 404)

Leroi-Gourhan's belief in the eventual demise of alphabetic writing and its illustrious products is expressed in the very medium whose disappearance it heralds. If his prophecy is successful, then any response to it from the archaic present will be inadequate and limited in ways expressible only by its successor. It also seems fantastical and infeasible—outside the weightless imaginings of speculative fiction—to think of something as deeply folded into our Western historical and religious being and cultural self-identity as alphabeticism disappearing. (Of course, from a non-alphabetic standpoint such as Chinese orthography, its demise might seem less impossible and more imaginable, but such a perception doesn't impinge on the question of archaization that prompts his prediction.) And yet is not the opposite belief to Leroi-Gourhan's just as outrageous? Can one really believe alphabetic writing will never be archaic, will always be with us, that in all possible, foreseeable, or imaginable futures of the human extending our present technologized state, alphabetic inscription will go on being the principal medium for recording, creating, and transmitting human knowledge, telling history, thinking philosophy, and inscribing affect?

Leroi-Gourhan asks us to imagine our present alphabetic graphic practices as ancient before their time. In an immediate empirical and manual sense, alphabetic writing already feels obsolescent: writing a text, such as the present one, by making half a million minutely different, attention-needing, error-prone, and irksomely intricate finger movements on a keyboard is, in the scheme of practical things, hardly less archaic than laboriously incising cuneiform syllables one by one into wet clay. But this seems not to be in the direction of his question. One might, mindful of the alphabet's limitations in respect to rendering the prosodic elements of spoken utterances, invoke the mark-up languages extensions to it being developed to remedy the situation. These so-called languages are scripts along the lines of an extended HTML embedded in text files; only instead of enabling a browser to display a hypertext page they enable voice-synthesizing devices to read aloud alphabetic texts. Their purpose is to improve the machine-readability of texts along affective dimensions by providing tags whose decoding allow text-recognition software to reproduce a passable version of certain standardized prosodic effects. But textual augmentation to improve machine vocality embeds us further rather than takes us out of the regime of alphabetic writing.

Of course it is possible that existing and not new graphic practices are in the process of rendering the alphabet archaic. Certainly the explosion of visual images brought about by digital technology has resulted in many of the traditional semiotic functions discharged by alphabetic writing (notably but not exclusively the display, recording, manipulation, and transmission of technical and scientific knowledge and information from demographic datasets to weather patterns) being usurped by visual artifacts—tables, graphs, arrays, diagrams, charts, and maps—replacing words; and it has resulted in changes in parellelist subject positions and subjectivity in directions quite distinct from that countenanced by alphabeticism and the textual protocols it furthers (more of which later). But again, notwithstanding the threat to alphabetic inscription, this is not what seems intended by Leroi-Gorhan's prognostication, since visual practices which have long performed their own dance in relation to alphabetic writing are governed as much by an orthogonality to written texts as by one of obsolescence or supersedence. After all, the rivalry between words and images is as ancient as the alphabet's acrophonic emergence from pictograms, and the iconophobia embedded in the alphabetic-text-based monotheisms of the West testifies to a long-established antagonism which would have to be dismantled or rendered irrelevant for the image to sufficiently destabilize the alphabetic word to the point of its archaization. What Leroi-Gourhan points to is not another move in an old battle between pictures and words but a new medium: a communicational technology which would operate in a manner related to the fact that the alphabet writes the movement of the speaking body.

The operative term here is *body*. For it is from an embodied ethnological perspective that he understands alphabeticism, writing being for him "the subordination of the hand to language," a servitude coming to an end as *Homo sapiens* is "freed from tools, gestures, muscles, from programming actions, from memory, freed from imagination by the perfection of the broadcasting media, freed from the animal world, the plant world, from cold, from microbes, from the unknown world of mountains and seas. . . ." (407) But forty years later, in the full flood of a digital transformation of our cultural infrastructure not suspected by Leroi-Gourhan or his contemporaries, we are beginning to understand that bodies and their (our) immersion in the world are not so easily abandoned, that escaping from corporeality merely replays an ancient fantasy of transcendence rather than follows a narrative of technological advance, and that contrary to withering the body or leaving it behind it will be by uniting with it—merging, augmenting, capturing, and re-engineering it—that technology might render our present alphabetic dispensation archaic.

In light of this, responding to Leroi-Gourhan's provocation requires attending to the link between alphabetic writing and the body, one which will go beyond, or behind, the usual instrumental and representational formulas governing that connection.

notating against capturing

The alphabet, as we've seen, notates a portion of the signifying sounds produced by the organs of speech. A suitably specific escape from its presumed archaic state might go by way of a double leap. First jump, beyond the written mark: why interpret 'writing' as notation, as the projection of body activity (here, speech) onto a pre-set list of inscribed marks and a syntax? Why not an a-symbolic mediation—a direct sampling or capture rather than a coded representation? Second jump, beyond the oral-vocal apparatus: why the restriction to the movements of the organs of speech, to the physiology and neurology of breathing and its articulation into consonants and vowels? Why not the movements of any and all of the body's organs and parts, oral, aural, or otherwise, traditionally signifying and a-signifying alike?

The dominating interpretation of notation and writing (outside the pragmatically unhelpful totalization of the latter as 'spacing' within deconstructionist discourse) is that of media comprised of marks and syntax with pre-assigned meanings ranging from the constative to the performative and operational. Within this formulation there are many different examples. Thus: the originating example, at least in the present context, is of course the alphabet whose letters (along with punctuation and other marks) are operational signals for a human reader to reproduce specified sounds and their relation according to a juxtapositional and linear syntax. Or the symbol system of Western five-line musical notation, where the sounds are to be reproduced by pre-calibrated voices and instruments and the syntax has a simple two-dimensional structure. And the various systems for notating dance movements, of which Laban notation with its larger number of symbols, inclusion of musical notation, and more elaborate two-dimensional syntax is the best known. And, older than alphabetic writing and in a category of its own, the vast field of mathematical writing with its open-ended array of ideograms and diagrammatic symbols and multidimensional syntax that can (and systematically does) become part of the meanings being mediated.

But against notational media are the practices, apparatuses, and modes of capture which constitute non-symbolic media. In these, what is mediated operates under the regime of the enacted or reproduced rather than the symbolized. Thus, the phonogram and tape recorder don't notate sound in the form of symbols but write it—record or capture it—as a direct signal to an apparatus able to reproduce (a perceptually indistinguishable version of) the captured sounds; likewise the camera for the reproduction of captured visual scenes (though with the additional layer of convention and subsequent interpretation of images not present with sound capture). Described in terms of figures of speech, notational media are regulated by metaphor and similitude, while capture media operate under the regimes of metonymy and synecdoche. Or, focusing on semiosis, notation involves a discrete algebraic framework or relational structure of prior terms, while capture presents a continuous topological model of posteriorly identifiable, internal differences.

The existence of media like photography, film, video, and sound recording, then, which are able to make a direct iterable trace of the look of the visual real, or the sound of the audible real, addresses the second of the two jumps just indicated by way of an immediate suggestion: why not an analogous form of mediation for the perception of the moving real.

Is there any reason—biological, physical, theoretical, or practical—why movement could not similarly be recorded? Could there not be a form of kinematic writing able to capture the perceptual feel of actual movements in space? The answer is yes, there could be such a medium: digital technology in fact offers the possibility of a non-notational medium capable of reproducing the kinesis of bodies, called, appropriately enough, motion capture.

But before elaborating, it will help to gloss common and for our purposes not irrelevant uses of the term 'capture' connecting it to older forms of mimesis. Thus, within traditional theatrical and ceremonial discourse one talks of imitation, mimicry, copying, and quotation as modes of capture, in which the human body and voice are used to capture movements, postures, and sounds of humans (but also of animals and machines). But these actions lack any trace or record of themselves, and though they function as vital forms of capture within artistic performance are not a form of writing as interpreted here—their impermanence, their lack of any iterable trace, being precisely the point of the epithet that theatre is 'written on the wind.' From the opposite direction there are those media which do indeed record and notate (unwittingly) a mimicked movement, though this aspect—which legitimates calling them written media—is not in their accepted description foregrounded about them. Examples would be painting, carving, calligraphy, pottery, sewing, embroidery, knitting, and weaving, all of which achieve a certain freezing of actions by reproducing traces of human movements (rather than the movements themselves) as inextricable components of the pots, written messages, depicted images, tapestries and textiles whose production is their primary function.

Of these modes of unintentional capture, painting by virtue of its cultural preeminence resulting from its representational usages is exemplary; its recording of the artist's gestures through brushstrokes constitute what is recognized to be an essential aspect of the art, an aspect, however, discursively downplayed and neglected when compared to painting's representational function. The art historian James Elkins, battling this neglect of the sheer materiality of paint and arguing against the marginalization of the embodied action and painted gestures in favor of the depicted content, insists on the paint-psyche connection, how paint "embeds thought," how it furnishes a "cast of the painter's movements, a portrait of the painter's body and thoughts." Seen thus, paintings, particularly oil paintings, become records of a thousand accumulated strokes of the brush, which "preserve the memory of tired bodies that made them, the quick jabs, the exhausted truces, the careful nourishing gestures." (1999, 5) Or, less deliberate and localizable, infamously brushless as it were, and detached from the depictive and representational agenda of figurative art, are the so-called action paintings of Jackson Pollock—paintings which enact a version of the corporeal axiomatic in that they embed the feelings, attitudes, and unconscious knowledge of the artist's entire body outside the orbit of any discursive or representational narrative. In the development of Western painting the resulting capture of body movements produced an entire style and language of painting—abstract expressionism—which rerouted the itinerary of modern painting. In the present terms, one can say that such expressionism is abstract precisely to the extent that concrete content has mutated from being a depiction of the object—which is precisely the domain of a notational medium—to the arena of capture in which content is encompassed trans-notationally as (the trace of) of performed movement.

Abstract expressionism exhausted itself long ago. It was eclipsed by performance art where its relation to the traced body of the artist gives way to the live mobile presence of his or her body, and more substantively by installation art in which there is a transfer of embodiment from artist to viewer as the viewing body is kinematically augmented by being required to move in, through, and out of the space created by the installation. But though the installation's overshadowing of the traditional flat image effectively terminated the function of the brushstroke as frozen movement, the presence of a mediated trace of the body's action didn't disappear but migrated to digital technology, where the potential for directly capturing gesture, as well as recording or tracing it in painting, has been given a new itinerary whose emergence from the original haptic device of the computer mouse was unintentional. Thus, as one knows, the mouse is the enabling technology of the Graphic User Interface; invented some thirty years ago as a move-and-click device for registering the user's (x, y)-position, it has in addition to this function since become a digital paintbrush, able to capture (traces of) the highly restricted movements and gestures of the hand swiveling at the wrist and sweeping across a rigidly delimited flat surface. The resulting brushstrokes can be used to create certain kinds of digital paintings—mousegrams one might call them—that exhibit these gestures, or rather reduced, metonymic projections of them, in a captured state.[6]

Preeminently among forms of painting, action painting captured traces of the artist's activities—imprints, casts, or residues of movement.

Whatever thought, affect, and feeling that painting was able to embed directly (non-representationally) into the painted surface via its pourings and brushstrokes had to be accomplished through these immobile reducts, what Châtelet calls diagrams, of the gestures and movements of the painter's body. Motion capture is able to escape from reducts and residues by being able to deliver a form of kinematic writing not confined to such stationary traces.

motion captured

One captures a body's motion[7] through the use of tracking technology, by attaching markers or sensors (responsive to visual or magnetic or aural or inertial tracking technologies) to chosen points on the body (of an animal, machine, human) and takes periodic readings, i.e., digitized samples, via cameras, magnetic sensors, and so on, of where in space these sensors are as the body moves. In fact, 'motion' as such is not essential; in the case of hapticity, the body might move (as in stroking) but does not have to; even potential movement, the propensity to move—pushes, pulls, impulsions, graspings—which issues in pressure, tensions, and dynamic resistance can be captured by force-feedback devices closely related to tracking sensors. In both cases, the resulting data-set of sampled readings contains the information needed to construct what can be perceived to be a reproduction of the original actual or potential motion of the chosen aspects of the body in an unlimited series of situations.

As the phonogram and camera enable the storage and production of sound and vision, motion-capture technology stores and reproduces kinesis. Two differences however should be noted. The first concerns historical development: unlike the capture of sound and vision, which were originally analog and only subsequent to their technical, cultural, and artistic establishment became digitized (allowing a contrast between the old analog content and its digitized form), motion capture leapfrogs this history by producing digital files from the outset. The second difference is operational: the haptic sense, unlike the more passive recording of sound and vision, has to be captured actively; observation is not enough. In order for the sense of touch to be registered some degree of enactment or manipulation must precede perception. But, notwithstanding the complications and creative facilitations these differences imply, there is a functional analogy between digital sound, digital video, and digital capture of kinesis, in that, like sound and video, the readings of motion-capture sensors are raw data, a metonymic sampling of their target motions, as opposed to the metaphoric and symbolic representation provided by all notational forms of writing.

The kinetic patterns stored by motion capture dis-embed, de-contextualize, and de-territorialize the original motion from the place, time, circumstances, physical form, cultural particularity, and presence of its performance. Released from their originating situations and instantiations, they can be re-territorialized onto a proliferating range of physical situations and re-embedded within any number of contexts unrelated to the original occurrence. Captured motion is able to be endlessly re-instanced and re-realized—to drive two- and three-dimensional animations, to effect the movement of an automaton, a puppet, a robot, a cartoon figure, an electronic doll, a virtual reality avatar, or indeed to become part of and enter into the movements and perceptions of another human body.

This latter has been exploited by the Australian body artist Stellarc, as part of his extended campaign to demonstrate that the "the body is obsolete" (by which he seems to mean the non-technologized, 'natural,' purely biological and unprosthetic body—were there to be such a thing), in a recent installation and experiment in which he sent readings of his arm movements via the Internet to drive the wired arm of a remote host, a person elsewhere on the planet, causing her to stroke her breast without any anticipated, conscious, or willed effort on her part. The result is a form of transposed corporeality: for the remote person it appears as a kind of automatism or possession from without, perhaps a virtual violation, while for Stellarc, or anybody else initiating it, it appears indistinguishable from a pseudo-masturbation, an indirect self-pleasuring enacted through another's body.

Once motion is digitally capturable it becomes digitally designable, giving rise to the possibility of motion generated in the absence of an original: capture by means of sampled readings and simulation by means of executed algorithms being two sides of the same process. Though still in the early developmental phase of the technology, captured and simulated gestures are already in widespread use, figuring in art objects, computer games, virtual choreography, animated films, different kinds of electronic installations, and various realizations of the concept of electronic or virtual theatre.

Likewise, though subject to different technological and cultural imperatives, captured forces operating via force-feedback devices are enabling varied forms of purely haptic action at a distance. These range from handling real and simulated molecules by research chemists to robotically executed tele-surgery controlled through visually enhanced feedback loops to cross-planetary arm-wrestling to the attempt to realize tele-dildonics—sex-at-a-distance which according to its adherents and would-be practitioners would allow a participant to feel his or her avatar sexing another's avatar, or as they put it "computer-mediated sexual interaction between the Virtual Reality presences of two humans." (Hornsby 2001) Differently and richer in the direction of more nuanced and interesting affectivity, force capture would include the full haptic modalities—all the numerous forms of stroking, squeezing, and touching—that organize the informal and largely uncharted inter-actions and corporeal communings assembled within our everyday constructing and enacting of human embodiment.

Motion capture technology, then, allows the communicational, instrumental, and affective traffic of the body in all its movements, openings, tensings, foldings, and rhythms into the orbit of 'writing.' There is in this no limit, at least in principle, to what about or of the body is capturable: the locomoting, cavorting, dancing, strutting, gesturing body; the arm, hand, torso, and head of the Signing body; the ever-present gesticulating body accompanying speech; facial expressions, transient twists, posturings, posings, and turns of the body in performance; shrugs, eye blinks, winks, barely perceptible tremors and nods of the socially communicating head and face; and so on. And not only the signifying and a-signifying moves and dynamics of people, but also of machines and animals and objects: from a chimpanzee's grin or the throbbing of a massage chair, to the movements of a musical conductor's baton, the swing of a golf club, the vibrations of a violinist's bow, and so on. Inversely, there will be actions and kinematic patterns that will exist purely by virtue of being generated and thus (until imitated) signifiable purely by virtue of this very technology. All of this constitutes a gesturo-haptic medium of vast, unrealized, and as yet untheorized or critically narrativized potential.[8]

And what, to revert to our original question, does this imply for alphabetic writing and for going beyond the writing of speech? And would not such in its turn go beyond the hegemony and authority (as this is presently constituted) of written discourse itself? Could motion capture be about to induce a transformation as radical and far reaching for the body's gestural activities, for its skin and organs of grasping and touching, as writing accomplished for the organs of speech? Could bringing (a digitally objectified) gesture out from under the shadow of the spoken word install a new order of body mediation? In such an installation it is not a question of the gesturo-haptic being equal to or even seriously rivaling the present-day importance and functional centrality of speech (which would be absurd), but rather the putting into place of an experiential, as opposed to linguistic and spoken, modality. Certainly, as we saw earlier, gesture's relation to speech is complex and many leveled: it can accompany and be intricately synchronized with speech (gesticulation); it can operate counter-orally and function to displace, nullify, or problematize speech (so-called emblem gestures); it can be a rival in the sense of an autonomous and complete alternate linguistic medium to speech (Signing); or it can operate inside speech as its armature and the vehicle for its extra-word affect (prosody). These gestural activities are evolutionarily old and have long augmented and conditioned speech on an automatic, everyday basis. This makes them ideal for designers of human-machine interfaces, since there is no difficult learning curve: "Communicative behaviors such as salutation and farewell, conversational turn-taking with interruptions, and referring to objects using pointing gestures are examples of protocols that all native speakers of a language already know how to perform and can thus be leveraged in an intelligent interface." (Cassell et al. 2001, 55)

Interface design is a pragmatic aspect of the gesture–speech nexus in which gesture is seen as augmenting and combining with speech. But, as indicated, gesture can be counterpoised to speech as a form of principled silence offering different ways of saying nothing; in this sense, what the gesturo-haptic amounts to is a new mediation of silence, a means of arriving at a new and productively positive valorization of a once contemptuously silent, dumb, a-rational, and emotional body. Or rather, a new, previously overlooked body in a constant state of arriving, since the saying of nothing, becoming mute, is inseparable from the never-ending business of creating a wordless interior to spoken language. To achieve the body without organs of speech (as Artaud might have put it), to dumb the body, de-organize it, divest it of utterance, so it no longer is governed by the sayable (and hence the writable) is to enable it to become the field of other productions, other desires, and to be alive to other semiotics, other mediations—here the gesturo-haptic—which speech, unable to process silence except as an absence of itself and yoked to its alphabetically written form that suffers from precisely the same inability, has always been only too pleased to elide.

Nobody indeed has argued with more passion against such elision and the corporeal truncation it rides on than Antonin Artaud. Precisely the opposition between the gesturo-haptic and the linguistic in relation to the means and protocols of theatrical performance was the overriding aesthetic justification and moral force of his theatre of cruelty. For Artaud the purpose of theatre was the induction of feeling, emotion, and affect, the bringing about of a physical and hence spiritual and metaphysical transformation of the bodies of the audience, a corporeal upheaval in its recipients whose effect on their subjectivity would be as discontinuous, violent, far reaching, and unignorable as the plague. Speech, especially the speech derived from a script, could not achieve this. Artaud, requiring his theatre to function with a "pure theatrical language which does without words, a language of signs, gestures, and attitudes . . . ," knew that speech could not but denigrate and marginalize gesture, whereas what was needed was a reversal of the dominion of thought over the body: "gesture . . . instead of serving as a decoration, an accompaniment of a thought, instead causes its movement, directs it, destroys it, or changes it completely" (1958, 39). Artaud's program of theatrical embodiment—his refusal to subordinate the stage to the performance of writing, his insistence on the imperatives and possibilities of screams, shouts, gestures, cries, primitive signals, inchoate utterances, and silences—did for modern theatre what Pollock's refusal of intentionality and conscious purpose and his insertion of the body onto the canvas did for painting: it reestablished the gesturo-haptic at the expense of the depictive, the representational, the linguistic, the textual, and the symbolic.

Becoming silent, acting mute is becoming infant, and as such is understandable as part of a willed accession to the state of pre-speech, a kind of return to or renegotiation of the past, except that what is involved in such a move is not a 'return' or any kind of regression, but a reconfiguration of the present by altering its genesis, its supposedly necessary and irrevocable linkage to that past by re-originating itself. The result would be an alteration in the conditions for the possibility of being human, a quasi- or neo-primitiveness in which future humans partook of the characteristics of (present-day) children. In this cultural neoteny whereby the adults come to resemble the young of their evolutionary forebears, speech would not of course disappear, but on the contrary become reconfigured (as it was once before when transformed by alphabetic writing), re-mediated and transfigured into a more mobile, expressive, and affective apparatus by the nascent gesturo-haptic resources emerging from the technologies of motion capture.

'writing'

Plainly the gesturo-haptic achieves both departures from the alphabetic inscription of the words spoken—jumping beyond notation and beyond speech—suggested earlier: its capture of the moving human body (not to mention animals and machines) far exceeds the alphabet's inscribing of the organs of speech. Nevertheless the gesturo-haptic presents itself as a form of writing or 'visual notation' (Furniss 2006) that bears a fundamental kinship with alphabetic writing by making available a set of effects parallel to the virtualizing action performed on speech by the alphabet. It certainly extends to gestures the same kinds of conceptual and pragmatic mobility, spatio-temporal dislocation, freedom from the contexts of their production, and analytic transparency as the notational system of alphabetic writing afforded human speech. For as we know, the alphabet, by allowing (insisting) that words become self-standing objects, discrete items of awareness that could be isolated, studied, compared, replicated, and systematized, gave rise to grammar, written discourse and literature, and a science of linguistics. Likewise, gestures with respect to their digitally captured forms: they too are now being identified, individualized, examined, replicated, and synthesized as discrete and autonomous objects of conscious attention. The opportunity is thus opened for such newly digitized and objectified gestures to emerge from the shadow of speech, to be 'grammaticalized' and give rise to a gesturology, the theoretical implications of which would extend to a re-thinking of the status of human corporeality. More practically, such a gesturology might serve as an enabling semiotic frame allowing the gesturo-haptic to function as a medium for Sign to possess what it has so far lacked—a stock of iterable gestural artifacts constituting a 'literature.' It might also do for the principled silences and unwords of the gesturo-haptic body, not least the production of that body's presence to itself and to others, what linguistics has done for spoken language on the level of positive knowledge, and what genealogies of discourse have striven to do for the constitution of the speaking subject.

But at the same time the gesturo-haptic is a form of writing which exceeds the textual, insofar as the 'text' can never be separated from the hermeneutic and the interpretative activities of deciphering. Its mode lies outside the text, the gram, or the trace of anything prior to its own performance, since it works through bodily enacted events and the necessity of being experienced as these occur. In this sense it is profoundly and inescapably exo-textual, a mediating technology that escapes the bounds of coded signification by operating within interactive, participatory, and immersive regimes. In other words, the gesturo-haptic doesn't communicate in the accepted sense—source A sends signifying item B to a recipient C— it doesn't convey messages, send information, transmit meanings, or bear significations which exist and are determined in advance of its action; what it traffics in are corporeal events in so-called real time, processes which have to happen, and in happening—better, in the manner of their happening—engender meaning. And in such a mode of mediation there is no pure B uncontaminated by an A and C, or vice versa, but rather a triangular interdependency of 'messages' and sending and receiving bodies. More conventionally: gestures (however isolatable they might be as discrete items of communication or objects of analysis) are not signs in Saussure's or Pierce's sense, except insofar as they become so retrospectively in that they signify (if that is the term) their own happening and its expected or habitual affects; their meaning in this retrospective semiotization is the fact and embodied consequences of their occurrence.

But, to repeat, however naturalizing this after-the-fact discursive description of the gesturo-haptic appears, it should not be allowed to mask the fundamental difference—the gap between language and experience, discourse and embodiment, representation and enactment—separating it from purely textual writing. Of captured items such as a handshake, a shrug, a squeeze of the shoulder, a kiss, a turning aside, and so on, one can say that despite their conventional meanings, what is salient about them—their importance, value, and strategic or instrumental interest—is not derived from these meanings, but in the fact of their taking place and in the subsequent psycho-social-corporeal effects (of affect, safety, assurance, threat, etc.) they induce and could only induce as a result of having occurred, and having done so in the manner, style, and force (all that constitutes what one might call their gestural prosody) that they did.

To think otherwise about gesturo-haptic mediation, to allow it to be reduced to a species of discourse and assume its effects to be wholly articulatable as such, is a double misapprehension, since it misperceives how the medium works through bodily transformation and not linguistic symbols and, less transparently, it obfuscates its direct—pre-discursive—action on the body, what I called earlier technology's corporeal axiomatic, and hence masks its action in inducing and installing subjectivity.

technologized subjectivity

"Cultures," Merlin Donald reminds us, "restructure the mind, not only in terms of its specific contents, which are obviously culture-bound, but also in terms of its fundamental neurological organization." (1991, 14) Likewise, as Terrence Deacon argues, a cultural phenomenon such as the development of language and its ramifications can be seen to have altered the size and overall capacity of the brain (rather than, as is usually supposed, the reverse). In Donald's case, the arena of this restructuring is culturally mediated systems of external memory (writing for example), an instance of technologically mediated exogenesis; as such it can be generalized from the memory-storage functions of written notation to their extension in the present account made available by the medium of gesturo-haptic writing.

A principal mode of exogenesis is synthetic assemblage, the coming- or putting-together of independent activities to form a new, functionally unified and autonomous entity with emergent properties not present in its components. Two evolutionary examples, far removed from motion capture, illustrate the point. One is human number sense. The ability to number things in a collection has been shown by Stanislas Dehaene (1997), contrary to the accepted view, not to be an endogenously formed faculty, but a capacity assembled by social demands (the imperatives of mathematical practice) from differently evolved and independent brain activities, each with their own functionalities having no prior intrinsic or necessary connection to number. The other is speech itself which though evidently a unified faculty (the supposed object of linguistics) is, as Deacon makes clear, the result of a coming together of many separate neurological capacities and physical changes from the propensity for vocal mimicry through left and right hemispheric specialization of the brain to the descent of the larynx enabling a sufficiently adequate range of vowel production (1997, 353).

The same synthesis enables a gesturo-haptic form of exogenesis. Captured gestures can become the elements for previously nonexistent and unknown assemblages of body movements, assemblages which are the site of neurological restructuring and in the presence of which new neurophysiologies, forms of corporeality, and subjectivities come into being. Going far beyond the neurological examples of number sense and speech, these assemblages will not be confined to the synthesis of elements drawn from internal and preexisting brain activities. Once the body's movements, gestures, and hapticities are captured and digitally manipulable they become, as we've seen, de-territorealized, which means they can be assembled outside the neurological confines of a (any) individual brain and put together through networks and culturally mediated collectivities, both existent and yet to be created. In this expanded field, captured body movements become the means of creating subjectivity—selves, subjects, and subject-positions—differently operative and differently sourced from those available within alphabetic writing. By allowing gesturo-haptic mediation this kind of constitutive role, such a rewriting of the corporeal provides a particular—technologically specific—concretization of Gilles Deleuze's paraphrase of Spinoza to the effect that "we do not know what a body is capable of," in that it reveals the body as both vehicle and recipient of becoming, as the site of a movement to and from the outside of the human, whose contemporary technological facilitation can be seen in the emergence of a bio-technic subject which is still in the stage of being characterized—dispersed, pluralized, de-centered, distributed, etc.—as negations of what it is in the process of superseding.

But however productive this sort of non-endogenous assemblage is, it presents only one strand—the explicit and manifest aspect—of the exogenesis brought about by technology. Technologies, as we've observed, restructure our neurology, to impinge on the body and its psychic envelope along specific channels: conventionally either as prosthetic extensions of physical, cognitive, and perceptual powers (the usual effects of tools, machines, apparatuses) or, as media, through the corporeal changes of affect and subjectivity wrought by the cultural products they make possible (the usual effects of the arts, literature, film, and so on). But less obvious and no less interesting, more so perhaps because they operate invisibly, are the non-explicit, unintentional, and pre-cultural corporeal effects of technologies: their recalibration of time and space, their facilitations of new modalities of self, and the work they do behind or beneath or despite the explicitly instrumental or signifying functions they are known by and are introduced to discharge.

This double technological restructuring operates at all levels from large-scale infrastructural phenomena such as electricity distribution or computer technology to the most banal machinic device. Thus, a trivial but paradigmatic example of the latter: the pop-up toaster is a tool invented to prosthetically extend the cooking body; less obviously, it contributes to the atomization of attention, micro-periodization of temporality, and intermittence of conscious awareness that are endemic to subjectivities operating in an electronically patterned infrastructure. Likewise, but in a diffuse and more indirect way, one we shall identify later, are the unintentional technologically driven changes in subjectivity associated with computational and visualization technologies (chapter 4).

Such effects are not moreover tied to hardware and physical apparatuses: the technology in question can be primarily cognitive, as is the
case with alphabetic writing. Here the explicit, intentional functions and
effects—prosthetic extension of speech, written discourse, the creation
of Western literacy together with its wider technological and intellectual
outworks—are evident enough, as are certain much commented upon
concomitant and collateral psycho-neurological effects of the alphabet,
such as the emphasis on linearity, the inculcation of analyticity, and the
promotion of context-free and atomized modes of thought. Less evident,
in fact, quite invisible (as to be expected, given technology's corporeal
axiomatic), is alphabetic writing's reconfiguration of the body at the level
of neurophysiology, an effect that installs a transcendental fissure or ontotheological resource inside its texts whose ultimate form is an abstract,
disembodied being—the God of Western monotheism. (See chapter 5.) In
light of this, to speak of the end of the alphabet is to suggest the possibility
of a shift in Western deism, a reconfiguration of God and the God-effect,
as momentous as the alphabet's inauguration of that Being. If this is so,
then the stakes for an end to the alphabet would be high indeed, and, to
return to Leroi-Gourhan's fantasy of post-alphabeticism we started from,
we have to wonder if such a thing is feasible; if an end to alphabetic writing or, less totally, a significant shrinkage in its universality, importance,
controlling functionality, and hegemonic status, is thinkable from within
that very writing here in the West?

Impossible to say, but Leroi-Gorhan's questioning of the ultimate
limits and end of the alphabetic text—like Marshall McLuhan's contem-
poraneous probing of the status of typographic man—is now being given
an extra twist by digital technology, a way of thinking the question in
terms of the transformation of the affective body and its subjectivities heralded by the yet to be realized mediological possibilities of gesturo-haptic
writing.

INTERLUDE

Three

TECHNOLOGIZED MATHEMATICS

A dominant theme of this essay is a tracking of the alphabet, its corporeal effects, its metaphysical and psychic legacy, and its mediological fate, that is, the possible end of the 'age of alphabetic graphism' in the face of technologies of the virtual. Part I laid out the alphabet's corporeal dimension vis-à-vis gesture and its textual elimination, and Part II will take up the question of its legacy and its ongoing displacement by digital media such as multidimensional visualization, parallel computing, techniques of simulation, and electronic networks—each of which is alien to the monadic, linear, and sequential protocols of alphabetic writing.

The effects of digital mediation are by no means confined to alphabetic writing. For example, the move from chemically bound marks on paper to weightless electronic notations on screens has already radically altered, revolutionized in fact, analogue inscription machines such as photography and mapmaking. And along a different route, computer-enabled simulation in the sciences—a virtual form of experimentation—is now recognized as a third investigative mode alongside the time-honored ones of theory and experiment. But, given that alphabetic writing is a discrete rather than analog inscription machine, the possibility exists that digital-based technologies of the virtual might consolidate rather than be antagonistic to it. In Part II we shall see this is far from the case.

One can consider another discrete writing machine, which like the alphabet operates through chemically bound marks on paper, namely mathematics, and ask what the effect of electronic writing might have on it—a question parallel to that which this essay is posing for alphabetic writing. The two machines, however, won't necessarily exhibit parallel responses. For one thing, simulation may indeed be a powerful new phenomenon, but mathematics, unlike the sciences, is already a form of virtual reasoning, a symbolic activity conducted solely through thought experiments on ideal, invisible objects. So that virtualizing mathematics further might exhibit paradoxical—de-virtualizing—effects quite different from anything suffered by the alphabet. For another, the digital computer that enables the whole electronic media scene is itself a mathematical object. What this means is that if the technologies of electronic visualization and simulation transform mathematics in any way, and I argue in this chapter that they do, they will do so in the form of a mediological feedback, a recursive loop whereby technologized outputs of mathematics alter the thought and practice of mathematics itself, a phenomenon that is surely not without interest. But focusing on the technologizing of mathematics, however interesting and relevant to the more abstract mediological aspects of this essay, is a departure from its principal, for the most part unmathematical, concerns, and readers averse to discussions about mathematics or for whom a little of the subject goes a long way can safely skip the following encounter with it and go to Part II.

Machines and Mathematics

We have seen how gesture is involved in the communication and creation of mathematical abstractions as a-linguistic models and sources of ideas, experienced as hunches, gut feelings, and 'intuition,' and as visible figures, frozen or arrested gestures manifest as mathematical diagrams. Another less direct link between gesture and mathematics is to machines. Certainly machines are closely related to diagrams which are often no more than material realizations of flow charts and blueprints—mechanical or electrical or digital-informational transductions of patterns of specific body movements and assemblages of gestures.

In fact, mathematics has been engaged in a two-way co-evolutionary traffic with machines since its inception. Mathematicians abstract concepts from machines—cyclicity and angular motion from the wheel-and-axle, ratios and rational numbers from the lever, modular arithmetic from clocks—and apply refined versions of them to create new machines, which are then the source of further abstraction, and so on. A materially framed historical account of mathematics alert to the presence and activity of the body would take this co-evolution as a datum. Such an account starting with the wheel and lever and then the pulley, the pump, abacus, clock, printing press, slide rule, punch-card loom, steam engine, camera, electric motor, typewriter, gramophone, radio, and computer would yield many particular versions of such two-way traffic. The interchange however will always appear asymmetrical: mathematics, augmented and permanently altered by its encounters with machines, sheds its connection to them when it presents itself as a formal, autonomous discipline, whereas machines in their designs and construction are never free of the mathematics behind them.

With the advent of the digital computer the entire co-evolutionary dynamic underwent a phase shift taking the machine and mathematics traffic to an entirely new, radically productive level. The digital computer was spawned by modern mathematics and metamathematics, and, with the exception of the abacus (its precursor), it is the most uncompromisingly mathematical of all machines ever, a realization of a mathematical abstraction—the Turing machine in the form of an electronic implementation of the binary logic and formal systems of mathematical logic.

My aim here is to suggest that the effect of the computer on mathematics (pure and applied) will be appropriately far-reaching and radical; that the computer will ultimately reconfigure the mathematical matrix from which it sprang and will do this not only by affecting changes in content and method over a wide mathematical terrain, but more importantly by altering the practice and perhaps the very conception we have of mathematics.

To justify such a claim we need a framework able to articulate the significance of the changes brought into play by the digital computer; we need, in other words, a general characterization or abstract model of what it means to do mathematics, to engage in the activity we call mathematics or, which comes to the same thing, what it means to be and function as a 'mathematician.'

A Model of Mathematical Activity

Behind the various construals of mathematics as an activity—investigation of abstract patterns, symbolic reasoning about ideal entities, study of number and space, and so on—lie three distinct, semiotico-practical discourses or domains that constitute the subject, namely idea, symbol, and procedure. By idea is meant the domain of human thought, mentation, and concepts from individual ideas to the most elaborate narratives and products of the imagination; by symbol is meant the domain of language, signs, symbols, and communicational and semiotic practices from notational devices and diagrams to entire linguistic systems; and by procedure is meant the domain of action, process, and operations from a single calculation, computation, or verification to an arbitrarily complex program of instructions.

Mathematics can be understood as a vast and particular complex of activities across these domains. The figure of the 'mathematician,' the one who carries out these activities, who knows and does mathematics, is on this understanding an assemblage of agencies, each of which implements the characteristic activity of a domain. So that, far from being a monolithic or indivisible actor, the mathematician can be seen as a plurality of parallel activities, a trio of different sources of action as follows.

The actor corresponding to the domain of the idea I call the Person. This agency has extra-mathematical physical and cultural presence, is immersed in natural language and the subjectivity it makes available, has insight and hunches, provides motivation for and is the source of intuitions behind concepts and proofs. Abstracted from the Person is the agency corresponding to the domain of the symbol, which I call the Subject. The Subject is the source of the inter-subjectivity embedded in mathematical languages, but is without the Person's capacity of self-reference, since it is circumscribed by languages which by definition lack indexical markers of time and physical place. Finally, there is the Agent, the actor associated with the domain of procedure who functions as the delegate for the Person through the mediation of the Subject; the Agent executes a mathematically idealized version of the actions imagined by the Person and it does so formally since it lacks the Subject's access to meaning and significance.

The model presented here, then, comprehends the mathematician as a tripartite being enacting certain kinds of inter-subjective thought experiments, imagined procedures conceived to take place in idealized times and spaces. The Person imagines an idea X associated with some procedure Y — the intuitive idea of number and the action of counting, for example. The Subject thinks X and Y within mathematical signs—for example, via a notation system which describes and determines what 'number' signifies mathematically and presupposes what is meant by 'counting.' The Agent, for whom X and Y are formal symbols and procedures, without meaning or significance, executes the action imagined by the Person—for example, the Agent carries out a calculation or other operation on a list of numerical expressions.

One of the activities performed by the Person is giving proofs of assertions. The Person makes a claim about an imagined task or procedure—

counting, inverting a matrix, etc. — that the Agent will execute; the task is articulated via an appropriate linguistic apparatus supplied by the Subject. Then, by tracking the Agent via this apparatus, that is by shadowing the Subject, the Person is persuaded as to the truth of the claim.

These three agencies are by no means absolute or fixed. This is obvious in the case of the Person who is manifestly a cultural and historical product. But it is also true of the Subject and the Agent: they are constructs. What constitutes them — the rules and protocols of legitimate sign usage, permitted definitions, legitimate procedures, agreements about valid reasoning and proof, and so on — have been shaped over time by innovation, conceptual reconfiguration, philosophical criticism, explicit programs of rigor, and the like. Having said this, however, one should note their relative stability: they do not change easily, and when they do they are obliged do so in relation to each other in order to preserve the coherence of the mathematician they jointly facilitate.

The latter constraint implies that a major change in one of these agencies is likely to have significant consequences for the others, and hence for the character of the mathematician and mathematics as a whole. This indeed looks to be the case at the present historical juncture. The computer is providing an Agent of incomparable power and flexibility, able to carry out procedures whose results were previously unsuspected. This frees up the Person to enlarge the scope and imaginative possibilities of mathematical thought, which in turn demands a Subject together with a suitably rich and powerful mathematical language for communicating between the Person and Agent. The result, as we shall see, is that a reconfigured triad of agencies able to operate in new and complementary ways is being put in place.

With this triadic model as framework we can return to the digital computer's effect on classical mathematics. In order to focus the question on essentials we shall restrict attention to what is essential and inescapable about the computer: it is a rule-governed machine; it is a digital as opposed to an analog machine; it is a material machine subject to and impinging on physical reality. The machinic aspect will result in local effects on mathematics involving reasoning and proof as well as the nature of iteration. The digital aspect will impinge on the notion of continuity which, in the form of the linear continuum, is the site where the analog is thought within classical mathematics. The material aspect will impinge on the supposed purity of pure mathematics, that is, on its separation from what mathematicians call the 'real world.' Since these notions form the corner-stone of post-Renaissance mathematics, it is likely that any change in their status will be significant.

The remarks about mathematics and computers that follow fall into two parts. The first is descriptive, outlining some of the varied ways the computer is currently altering the logical status, content, practice, and evolution of mathematics, the main thrust of which is a displacement of the infinite and the continuous by the discrete and finite along with the openings such displacement bring into play. The second part is more discursive, examining a sense in which infinity might be a philosophical or theoretical problem for computer science. Specifically, I look at a contention by cryptographers that computer science's infinity-based idealization of a 'feasible' algorithm is irrelevant to practical, real-world, real-time computational tasks.

Local and Global Transformation

local effects: machine reasoning

As indicated, every machine will have natural and expected local connections with the mathematics contiguous to it, that is those mathematical objects, practices, structures, and topics it naturally requires and facilitates and in some cases invents. For the lever it might have been rational numbers; for clocks and calendars, counting and the arithmetic of repetition; for the pendulum, the concept of a differential equation; and so on. For the computer the mathematics natural to it ranges widely since it embraces two core aspects of the subject: the computational and calculational process itself and the notion of proof in the form of mathematical logic and metamathematics.

The local effects that flow from the computational process have to do with mathematical content, with kinds of mathematics that are inspired by, demand, are amenable to, or are directly fostered by, computational and calculational procedures. Many areas of mathematics are covered by such a description, including much of graph theory, finite combinatorics, cryptography, and the entire theory and practice of cellular automata (of which more below); in addition there are fields such as chaos theory, nonlinear dynamics and fractal geometry, which pre-existed computational methods but only flourished with the entry into mathematics of the computer and which are now inseparable from the computational processes permeating them. And, more at the periphery of mainstream mathematics, there are those areas such as the investigation of symbolic systems, formal and artificial languages, the theory of algorithms, and finite model theory which, whilst not computer driven, relate naturally to the investigation of computational questions. Indeed, a subfield of one of these—that of algorithmic complexity—gave rise to the P = NP question, now considered one of the major open problems of contemporary mathematics. We shall examine an aspect of this problem below.

The local effects associated with the logical, metamathematical aspect of computers are oriented not to content but to method: to questions of reasoning and syntax, modes of mathematical argument, deduction, inference, proof and validation, and the mechanics of formal systems. One form this has taken has been the development of computer-generated proofs, computer-aided reasoning, and theorem proving by machine. In a sense there is nothing new to such an enterprise. Ever since Pascal's invention of a calculating machine and Leibniz's celebrated call to replace all disputation and rational argument by formal, symbolic means, there have been numerous and varied efforts to mechanize reasoning. The contemporary versions of these range from having the computer find new proofs of existing theorems to the more impressive goal of generating proofs or disproofs of open questions as part of efforts to rival traditional, entirely symbolic methods of mathematical proof and deduction by computational validation. Some of these have yielded positive results; one such theorem-proving procedure successfully settled a question, the so-called Robbins conjecture about certain algebras that had been open for sixty years. (McClure, 1997)

Less ambitious is the use of machine reasoning to augment rather than replace the traditional symbolic modes of mathematical argument. The main need for augmentation is the difficulty of managing large volumes of data and information involved in the handling of cases, circumstances, and effects so numerous and heterogeneous as to put them beyond the capacity of an individual mathematician's cognitive wherewithal, not to say lifetime, to examine. For example, the theorems underlying the classification of all simple finite groups relies on a computer-compiled database too massive and unwieldy and internally complex for a single mathematician to process. (Soloman, 1995) Though ostensibly more mundane than using computers to prove theorems outright, this augmentative approach raises a fundamental issue as to the nature and assumed constitution of 'the mathematician.' A famous example is the four-color problem which asked whether every map drawn in the plane could be colored using only four colors. This was settled positively by Kenneth Appel and Wolfgang Haken in 1976 by a proof which relied on a computer program to check thousands of intricate map configurations too complex to be verified in any other way. (Appel and Haken 1989) A different kind of departure from traditional methods of logical validation made possible by the computer occurs with so-called probabilistic proof procedures which run a series of spot checks on randomly selected fragments of long proofs and estimate to a given degree of confidence the correctness of the whole. (Goldreich, 1995) The question raised by probabilistic proofs, by the example of fi-nite groups, and more pointedly by the Appel-Haken argument is simple and profound: what is or should count as a proof? Traditionally, a proof had always been a persuasive argument, a logical narrative which could be checked and assented to, step by step, by an (in principle, any) indi-vidual mathematician. Evidently, such is not possible for the collection of all simple finite groups nor for the ensemble of maps that arise and need to be examined in relation to coloring. Plainly, these examples of augmented proof call for the idea of 'the mathematician' to be made explicit in a way this has not been necessary hitherto. Specifically, two issues arise: 'who' or what are mathematical arguments addressed to? and how is this addressee to be 'persuaded' by a putative proof? In other words, what is or should be the constitution of the Subject, i.e., what cognitive capacities and technologies of inscription and representation should be available to it? How — according to what principles — might persuasion take place, once it is mediated by a computer program and so removed from the familiar (but in truth largely unexamined) orbit of an individual assenting to logi-cal steps? Both these questions are about the couplings between the triad of agencies constituting the 'mathematician,' with the second requiring the idea of persuasion as a dynamic involving the Person and Agent to be made explicit. (Rotman 1993)

global effects: simulation not proof

Local effects, then, represent computer-inflected changes in mathematical content and forms of proof. In terms of our model of mathematical ac-tivity, they are principally the concern of the mathematical Subject and its relation to the computational Agent. More far-reaching are the epistemic changes involving imagining, understanding, conceiving analogies, and having intuitions, insight, and hunches which concern the character of the Person and its relation to the Agent. However, the chief effect here — that of digital simulation — reaches beyond mathematics to cover the entire field of contemporary technological and scientifico-mathematical practice and is so widely operative and productive that it has been hailed as an entirely new, third method of scientific research, a twentieth-century, specifically computational mode different from and not reducible to the two existing research modes—the experimental and observational that founded science in the seventeenth century and the much older theoretical and deductive mode basic to mathematical reasoning. (Kaufmann and Smarr 1993)

To describe the effects and scope of digital simulation it will be useful to note mathematics' relation to physics. From the time of Galileo until well into the twentieth century the bulk of external problems and questions put to mathematics came from the natural sciences, principally physics. This coupled with assumptions within physics of the smoothness and continuity of nature, inseparable from the infinite divisibility of space and time, legitimated and fostered a mathematics dominated by an infinite continuum of real numbers, infinite series, and the differential equations of calculus allied to these notions. But in order to operate at all, computers need to discretize calculus by converting these equations into finite difference equations. But even without the effects of such discretization, physics' employment of infinitary, continuum-based mathematics is no longer secure. Well before the advent of the computer, quantum physics was born on the overthrow of the assumption of infinite divisibility of the quantity physicists call 'action' and its replacement by the formalism of discrete quanta. Moreover, in empirical terms physics is not merely non-infinitistic but boundedly finite: such easily named quantities as 10^{-100} meters or 10^{-100} seconds or 10^{100} particles or 10^{100} light years have no physical meaning in the universe we inhabit. Since the introduction of computational methods the move against continuum-based mathematics has strengthened: various projects over the past three decades have argued why and how physics needs to be 'finitized.' (Greenspan 1973; Feynman 1982; Wheeler 1988; Landauer 1991). One of the most extreme and uncompromising finitization projects is Edward Fredkin's hypothesis of "Finite Nature," according to which "at some scale, space and time are discrete and that the number of possible states of every finite volume of space-time is finite." (Fredkin 1992) Along with this goes the insistence that the entire physical universe be understood—somehow—as a single device, a digital computing machine whose computations are the processes of the universe and whose software constitutes what we call the laws of physics.

With this in mind let us return to simulation—by which is meant a digital mock-up or electronic model of a given physical or mathematical process. To mimic a process by computation it is necessary to create a digital version of a typical state of the process together with a rule for changing states with the idea that iterating the rule through repeated recursion will give a simulation of the original process. To witness the mimicry the results are visualized, that is plotted, graphed, or displayed so they present a picturable world, a visual, on-screen image whose electronic behavior can be software manipulated. A simulation is created, then, in two phases, discretization and visualization, which together constitute a virtual world close enough to the original one (physical or mathematical) to be a valuable investigative and creative tool.

All of the above applies to mathematical as well as physical worlds. However, in the case of mathematics, where there is apparently no external world to be simulated and which is already in the business of creating, representing, and studying virtual worlds, the importance and novelty of the simulation mode is easily missed. One reason is that the visualization it offers seems merely an extension (enormously refined and powerful to be sure) of time-honored methods of plotting and curve drawing which, however useful as cognitive aids, are eliminable—adding nothing substantive to mathematics. Such a viewpoint, unproblematic for most mathematicians, is misleading for several reasons.

In the first place the very premise that diagram and curve drawing in mathematics are epiphenomenal, of undoubted psychological and practical value but of no epistemological worth, is a mistaken result of programs of rigor, such as logicism or that of Bourbaki's rewriting of mathematics within a set theoretical framework, that neither know nor are interested in how mathematics is actually made or changes.[1] Secondly, plotting and visualization techniques in the past were severely limited in scope and epistemic importance by the difficulty of performing more than a small number of the relevant calculations, a fact which made it easy to naturalize their results into a mere picturing of preexisting ideas, to be jettisoned once the idea had been symbolized. Quite to the contrary, the computer creates complex topological surfaces, fractal functions, and iteration-based entities which were previously not only invisible but unimagined, unconceived. The consequent increase in mathematical knowledge has been considerable, ranging from the discovery of new topological surfaces to the opening up of novel research topics. (Friedhoff and Benzon 1989; Hanson et al. 1995) Clearly, it would be a reductive misapprehension to think of simulation as 'mere' visualization. Thirdly, the very focus on the visualization phase allows digital simulation to be reduced to the business of picturing and so occludes the radically transformative changes set in train by the prior phase of discretization and the real-time iteration of a rule it facilitates—both of which engage the finite discretum and not the infinite continuum.

But, leaving visualization aside, the style and character of a simulation-inflected mathematical science—pragmatic, material, experimental—breaks with mathematics' traditional understanding of itself as a pure theoretical and deductive science. The advent of an electronically modified Agent destabilizes the triadic assemblage of actors that has held the pure 'mathematician' in place since classical times. In the wake of this empirical—that is, impure—Agent comes a digitally refigured Person who can intuit, imagine, and recognize the new sorts of mathematical objects, simulations, and iterations, delivered by computation; and at the same time a digitally refigured Subject is required with the language and writing system—the appropriate mathematical 'software'—necessary to mediate the traffic between Person and Agent.

Traditional mathematics proves the existence of its abstract objects and establishes claims about them by constructing logically controlled narratives of imagined procedures using an Agent who is an ideal, infinite possibility. Simulation mathematics substitutes empirical observation for logical validation and exhibits rather than proves the existence of its material objects via actual, real-time computations; its Agent is a material, finite actuality.

What, then, are the consequences for future mathematics of this difference? As we've already observed, there will be a change in practical methods of mathematical enquiry: empirical mathematics frees the Person of the burden of proof and thereby liberates other, previously uninteresting or unrecognized ways of mathematical thought and activity. This has implications for the different kinds of objects and processes that the two styles can engage. Essentially, mathematical assertions are predictions about what will happen if certain specified operations are carried out. By insisting that its assertions be logically proved, classical pure mathematics restricts itself to investigating procedures whose outcomes, in principle at least, can be predicted. It is not equipped to investigate the fate of uncompressible processes whose outcomes cannot be foretold; programs that have to be run in order to reveal their final state; sequences of os and 1s which are only specifiable by a rule equal in length to them. These are precisely the sort of mathematical object simulations empirical mathematics handles well. And since there are in fact vastly more of such uncompressible processes and irreducible sequences than the predictable kind, the territory of empirical mathematics is potentially huge compared to that of the classical variety. This is essentially the claim Stephen Wolfram makes for his "new science" based on simulation methods (computational iteration of simple programs such as cellular automata) which represents a "major generalization of mathematics—with new ideas and methods, and a vast new areas to be explored." (2002) Finally, to put the matter ontologically, in terms of what exists and is observable mathematically by virtue of the very materiality of computation, we can note the importance of actual versus ideal counting and iteration procedures: simulation mathematics achieves its difference and power precisely because it has access to effects—the actual results of iterating real-world operation—unavailable to the idealized, necessarily internal, that is rigorously separate from the world, regularities of classical reasoning.

Infinity inside Computers?

an intractable problem

As we've observed, for a physics increasingly committed to a discrete, quantized universe, the use of infinity within its mathematical apparatus— via continuity, unlimited divisibility, and real numbers—raises certain problems. For theoretical computer science, such a consideration seems irrelevant, since its objects and processes and their mathematical formulations are discrete from the start.

But this doesn't settle the question of theoretical computer science's relation to infinity: the science rests on the integers which form an infinite progression, and this very attribute seems as intrinsic to computing as infinite divisibility was until recently to physics. In light of this, one can legitimately ask whether the presence of infinity within its mathematical formalism is, or might turn out to be, a problem for the science of computing.

I shall respond by concentrating on a combinatorial difficulty that goes back to the early days of the subject, which has subsequently given rise to a mathematized and highly generalized meta-problem, a problem about problems, namely the notoriously difficult P = NP question, "considered to be one of the most important problems in contemporary mathematics and theoretical computer science." (Sipser 1992)

A famous example of the difficulty that P = NP mathematizes is the Traveling Salesman Problem. A salesman has to visit each of N cities just once and return to his starting city. He has a table of all the intercity distances and wants to calculate the length of the shortest circuit that will accomplish this. How should he proceed? The problem seems ideal for the mindless number-crunching computers excel at: the salesman tells the computer to find the length of every circuit in turn and output the length of the shortest. But this involves an extravagant computational cost as N increases. Thus, for small values, say N = 10, the number of circuits to check is $(N - 1)! = 9!$ or 362,880 circuits, a computational fleabite for present-day computers. For a slightly bigger value, say N = 20, the number is approximately 10^{17}, a very large collection of circuits but checkable in a few months by an array of state-of-the-art computers. For N = 50, or so, no present-day computer or any conceivable improvement thereof could complete a search through all possible circuits, whilst for N = 100 the number is outside the realm of the possible.

Though it features distances and circuits through cities the underlying issue here is not tied to these notions. The difficulty facing the Salesman surfaces in a vast web of seemingly unrelated problems, numbering several thousand, that arise naturally in graph theory, scheduling, routing, and timetabling, in cryptography, in Boolean logic, and so on over a wide range of disparate combinatorial situations. Some, like the Salesman's question, are optimization problems: they ask for the largest or smallest value of a numerical parameter such as circuit length, flow through a network, the size of a subset, etc. Others are in the form of decision problems which simply demand a yes or no answer, such as the Boolean Satisfaction Problem: given a Boolean expression in N variables, is there an assignment of 1 (Truth) or 0 (False) to the variables such that the whole expression has value 1? The two forms are for the most part convertible into each other.

Regardless of their diverse contexts and origins, all these problems can be seen as posing the same basic question: Is there a way of finding the optimum value or the yes/no answer that does not resort to the brute-force method of checking through all the possibilities? In computational terms, is there an algorithm accomplishing the relevant task that runs in a feasible amount of time, one that doesn't, in other words, require an impossibly large number of computational steps for a small or reasonably small value of its input variable N.

To make this question meaningful, theoretical computer science requires a mathematical definition of 'feasible.' Now, one cannot expect such a definition to address particular numerical facts (the computational leap from N = 10 to N = 50 we saw above) or, by the same token, to address what 'reasonably small' might mean. As a consequence, the question of feasibility is perforce posed generically, in terms of an unrestricted number variable N, and takes the form of a comparison between one type of mathematical function of N and another for all values of N. Thus, since brute force requires an exponential number of computational steps ([N − 1]! for the salesman's problem), one has to define 'feasible' to mean a number of steps bounded above by some function f(N) which must be guaranteed to be smaller than any exponential function of N. The decision by the computer science community to choose f to be a polynomial function was very natural: polynomials are a robust class of functions, closed with respect to addition, multiplication, and composition whose well-known properties include the crucially relevant mathematical fact that any polynomial function is eventually smaller than any exponential function.

Defining feasible as polynomial time was the founding move of the theory of algorithms, and though it attracted certain criticisms, the choice made possible the development of complexity theory, an "elegant and useful theory that says something meaningful about practical computation." (Papadimitriou 1994) Elegant certainly, but just how meaningfully related to practical computation is, as we shall see, a matter of dispute.

For theoretical purposes the yes/no decision problem formulation rather than the optimization formulation is more convenient. The class of all decision problems solvable in polynomial time is denoted by P. Observe that if a problem D belongs to P, that is, if there exists a feasible algorithm finding a solution s answering D, then D has the additional property that any proposed solution s to D can be checked for correctness in polynomial time. This is because one can run the algorithm solving D and compare the result with s to determine whether s is or is not a solution to D, all of which requires no more than polynomial time. The class of all such decision problems which can be checked in polynomial time is denoted by NP. Our observation amounts to the fact that P is automatically a subclass of NP. The great open question enshrined in the equation P = NP thus asks for the reverse inclusion: Is NP contained in P? In other words, if one can *verify* a possible solution to a decision problem D in polynomial time, can one *find* a solution to D in polynomial time?[2]

Evidently, the concept of polynomial is the organizing idea behind the P = NP problematic at every level; without it the question would collapse. This is so for several reasons. Firstly, the problem itself, as a mathematization of specific combinatorial situations such as the Traveling Salesman's problem, depends totally on the polynomial-based definition of feasibility;

in fact the problem is an artifact of this definition, brought into being with it and impossible to conceive without it. Secondly, its encapsulation of the many thousands of diverse concrete instances, of which the Salesman's problem is merely one, into a single meta-problem depends repeatedly on the unbounded freedom that results from polynomial composition, namely if f( ) and g( ) are polynomial functions then so is f(g( )). Last, and by no means least, the formulation would make no sense in terms of its thousands of motivating concrete instances, all of which are resolvable by brute—exponential—search routines, were it not for the mathematical fact that any polynomial function is smaller than any exponential function.

But it is this last, quite crucial, fact that is precisely the site of a dissatisfaction with complexity theory, the point at which its practical relevance has been challenged; and not coincidentally, as we shall now see, it is where infinity enters the mathematical understanding of algorithms put in place by computer science.

asymptotic growth

Defining a feasible computation as one with a polynomial run time rests, then, on a basic inequality—that polynomial functions are smaller than exponential ones. The guarantee behind the inequality lies in an asymptotic definition of 'smaller than' which refers to behavior of functions 'at infinity.' Specifically, a function p( ) is (asymptotically) smaller than a function q( ) if there exists some number M such that $p(N) < q(N)$ for all values of N greater than M. In other words, the definition refers, in a way that cannot be eliminated, to sufficiently large N. Thus, only in the limit, as N goes to infinity, is the relative size of two functions resolvable, so as to make it true that any polynomial function p is smaller than any exponential function q.

But suppose allowing passage to the limit, that is, invoking sufficiently large values of N, is ruled out? What if the values of N are constrained to lie below some fixed bound b? In that case, the guarantee vanishes, however large b is: one can, in other words, no longer assume, in the presence of a bound, that polynomials are automatically smaller than exponentials. In fact, under the condition of boundedness, the reverse can always be made to happen, polynomials can be larger than exponentials: for any exponential function q( ), it is always possible to find a polynomial p( ) such that $p(N) > q(N)$ for all N below b.

The definition of feasibility is supposed to ensure achievable computation times. But what, in the real, finite world where the demand for it originates, is to be meant by achievable? For, on any presently conceivable understanding of physical reality the existence of a bound b is inescapable. This will follow from local limits, such as those imposed by computer scientists not being able to wait for unspecified periods for algorithms to terminate, to global ones, such as the universe we inhabit appearing not to permit an arbitrarily large sequence of physical actions to be executed.

The existence of a bound obviously disallows any appeal to an infinite range of values of N and so prevents the theory of complexity from operating. Indeed, the presence of a bound brings into relief a clutch of questions: What sense does it make for a science of computing to formulate its fundamental concept of feasibility in terms of an infinitistic criterion that necessarily appeals to numbers outside the reach of its computational processes? Pragmatically, what is the relevance of the P = NP question for categorizing (let alone understanding) the possibilities and limits of computation? Why would its solution have anything to say about real feasibility, about the possibility of finding or not being able to find algorithms that worked within the limits and freedoms of the real world rather than those of the idealized, boundless universe of infinitistic mathematics?

These questions are far from idle. In cryptography, for example, one has to ensure that an encrypted message is safe from attack, that no method of decryption with a practically achievable running time exists, regardless of how such a method might be mathematically characterized. In such a context, equating 'practically achievable' with the existence of a polynomial-time algorithm makes little sense, and the entire P = NP problematic begins to appear beside the point. Unsurprisingly, this is precisely the judgment voiced by many cryptographers and engineers more concerned with practicalities than mathematical elegance. Thus, for example, about the central cryptographic task of finding the factors of an integer, Steve Morgan observes, "Even if factoring is polynomial, it isn't necessarily practical. . . . A polynomial algorithm of say the order of N^{20} is essentially intractable even for small values of N. Conversely, an algorithm that ran in the order of $2^{(0.1N)}$ would make factoring billion bit numbers easy." (quoted in Reinhold 1995)

Moreover, it is not even necessary to invoke high-degree polynomials to make the point; exactly the same objection arises from the possibility of large multiplicative constants being present. Thus if k is such a constant, 10^{30} say, then an algorithm running in linear time, that is of order kN, will be intractable for small values of N, whereas an exponential algorithm of the order of $2^{(N/K)}$ would be tractable for all values of N that could possibly occur in practice.

These examples illustrate that not only is excluding exponential running times and identifying feasibility with polynomial time not to the point, but the identification runs counter to the underlying intuition relating to 'smallness,' namely that if (the input to) a problem is small, one wants a feasible computation solving the problem to be small or at least near small. But while 10, say, is by common agreement a small value of N, one could not claim that 10^{20} (one hundred billion billion), let alone 10^{30}, is small or nearly so. These objections have been long known and long shrugged off by complexity theorists as the price of doing (asymptotic) business. If prodded the only response they can offer is that N^{20}, though undeniably a polynomial function, is artificial, that in practice, most familiar algorithms are of the order N^n where n is possibly as large as 5 or 6 but certainly a number much smaller than 20, and that likewise large constants such as k instanced above are simply not natural. But this appeal is hardly in good faith: the concepts of 'natural,' and 'in practice,' as well as a notion of 'scale' or 'size,' and the opposition of 'small'/'large,' are precisely what have been expunged from the theory's asymptotic account of feasibility.

In fact, the P = NP question, and along with it the entire apparatus of algorithmic complexity and the asymptotic characterization of 'feasible' that engendered it, is not merely irrelevant to cryptography, but has been deemed conceptually deceptive. For Arnold Reinhold, the apparatus is a "naked emperor," whose effect has been that "a whole generation of computer scientists wrongly believes that complexity theory illuminates computation and that the P = NP problem is the missing link to a theoretical basis for cryptography." (1995)

Nor is it the case that cryptographers have special, ungeneralizable needs. Their criticism is sharpened perhaps by the financial, security, and military stakes involved in questions of decryption, but their overall objection stems from unavoidable pragmatic demands; as such it arises in any technological or engineering arena that looks to computer science for algorithms to solve real-world numerical problems in real-world times.

The presence of a bound, then, not only eliminates the possibility of any asymptotically based, and in particular polynomial, criterion of feasibility, but it offers a basis to start rethinking the question of feasible computation. Thus, as has been suggested by several researchers, one can divide the integers into two separate regions, H1 and H2, below and above b respectively, and work inside H1 within a limited universe of discourse situated far below b. Thus, Reinhold, for example, suggesting the value $2^{\wedge}(2^{1000})$ for b, goes on to comment that "the undue attention paid to classical complexity theory arises from an inclination to assume results that are true in H2 must somehow be true in H1. But there is neither a theoretical not practical basis for this belief." (1995)

Indeed, there is not, and working below b takes a first step toward rethinking the feasibility question. It achieves this by making explicit in a directly practical way the counterproductive nature of 'letting N go to infinity.' Taking b as a dividing point induces a numerical scale or size into the proceedings and hence the possibility of defining 'small' and 'large' in relation to it. But as it stands the approach is crude: the bound b is imposed from outside, and its value, chosen deliberately to be so large as to be remote from any actualizable computation, is for this very reason unconnected to anything real or observable. Any scale thus induced will be extrinsic to such computations and their limits and unable to state (let alone address) the expectation, attached to the intuitive notion of feasibility, that computational time of a solution be of the same 'order of magnitude' ('league,' 'ballpark,' 'size,' 'numerical reach') as the problem; that is, the expectation that for 'small' N the solution time of an N size problem should also be small or nearly so but not in any event 'large.'

The irrelevance to cryptography of the P = NP question foregrounds a disconnect between an infinitistic mathematical concept of feasibility and an intuitive, pragmatic one, a gulf between the demands of real-world computation and the idealized model associated with the classical sequence of integers. It will be obvious that the gulf is a re-occurrence, posed in a particularly dramatic and focused form, of the theoretical difference between empirical and classical mathematics identified earlier. It would seem natural, then, to attempt to tackle the question of feasibility and hence perhaps the P = NP problematic anew using the resources and possibilities put into play by simulation methods. Whether this attempt to use empirical mathematics can yield dividends is surely knowable only by experiment, but it is not clear that the attempt makes much sense as an approach to the P = NP problem itself: the very status of this question as a preeminent unsolved problem of contemporary pure mathematics indicates how deeply folded it is within the worldview of classical mathematics. In which case the N = NP problem might be the baby that empirical mathematics throws out along with the infinitistic bath water.

Granting this, is there a different mathematical framework for thinking 'feasibility,' one which tries to navigate between the classical and empirical routes? If there is, finding it would seem to entail overcoming a large and difficult obstacle: how to effect a conceptual escape from the great attractor of the classical integers. No easy matter, since it necessitates refusing the core belief of classical mathematics, the time-honored "Dogma of Natural Numbers" (Rashevskii 1973), namely their 'naturality,' their objective, non-human, existence as the 'given,' 'true,' and only idealization of actual counting and iteration.

The integers embody a formalism and an idea, a syntax and a semantics; one can identify two sorts of attempts to escape from this dogma depending on which aspect is given priority. The first makes questions of significance and meaning subsidiary by starting from a symbolic apparatus based on formal definitions and concepts able to be worked to produce mathematical results, and only subsequently does it evaluate the meanings thereby facilitated and imposed. The second reverses the procedure and insists on an intuition first, a convincing alternative picture of what counting and iteration are to mean before any particular formalism is offered. Examples of both can be found. The first approach has been pursued by Vladimir Sazonov who introduces the inequality loglog N < 10 as a formal axiom constraining the value of any N considered to be a feasible number, and proves various metamathematical theorems on the internal consistency of the effect of such a definition (Sazonov 1995, 1998). The second has been pursued by the present writer who focuses on the semiotics of the process of repetition inherent to the integers to refuse their 'naturality,' and outlines an alternative, non-classical understanding of counting and arithmetic (Rotman 1993, 1998). Whether either of these approaches will succeed in overcoming the dogma of their naturality remains to be seen; evidently, the stakes in challenging the notion are high.

Conclusion

Will the computer transform classical mathematics? Surely this is no longer in question. And it is reasonable to predict that it will do so along the same dimensions—machinic, digital, material—which have determined the changes it has so far wrought. Let me summarize.

The computer is a machine for investigating mathematical reality; it is reconfiguring the mathematical imaginary and mathematical 'reasoning' in relation to repetition as radically as the microscope and telescope reconfigured vision and 'seeing' in relation to scale; in its wake mathematical thought will never be the same. Its machinic ability to recursively repeat the application of a rule far beyond unaided human capacity enables it to transcend the knowledge and proof available to such capacity, principally by engendering hitherto unknown and unsuspected mathematical objects, thus confirming the fact that repetition of the same can produce difference. In this it appears to replay—on a fantastically more complex plane—the primal creative act of mathematics, namely counting, which produces an endless stream of new, ever-different numbers through the recursive iteration of the selfsame operation of adding the unit.

The computer is a machine whose conception, operation, construction, and infrastructure are digital: it is antagonistic to the analog and to notions of smoothness and unbroken continuity associated with it. As a result it presents a conceptual and methodological opposition to the real-number continuum which is the formalization within classical mathematics of these notions, and, less directly, to the concept of infinity folded into the very concept of real number. A major effect of this opposition to continuum-based mathematics, aided by independent moves within physics to quantize all the parameters measuring physical reality, as we've seen, is to push mathematics away from infinite methods in the direction of the discretely finite.

The computer is a material machine as against the ideal, immaterial object of classical mathematics, the Turing machine, on which it is modeled. Two consequences flow from this fact. One, the topic of part two of the present essay, concerns a difficulty that arises when this difference is elided and the ideal, potentially infinite mathematics of the model is used to describe and investigate real-world computations. How seriously theoretical computer science takes this difficulty remains to be seen. The second consequence, more pertinent to the future of mathematics, stems from a kind of reversal of the first, namely capitalizing on the difference between real and imagined counting. The reversal works by actually running a program on a material machine and observing the result rather than theoretically running it on an ideal machine and predicting the result, a procedure which delivers into mathematics effects from the so-called real world, effects—mathematical objects and relations—available to the mathematical Person in no other way. The result is the opening up of mathematics by a powerful new Agent in the direction of an empirical science; a development that not only goes against its time-honored status as a purely theoretical pursuit, but is a dramatic and radical reversal of the last two centuries' efforts to eliminate all traces of the physical world from mathematics' definitions and methods in the name of an abstract program of mathematical 'rigor.'

Finally, a theme of this chapter has been the computer's impact on classical mathematics' core concept of infinity. This should not be misunderstood. Certainly, the computer's ongoing colonization of mathematics, particularly in the direction of empirical simulations, is bringing about a widespread de-emphasis of the concept. But to say this is not to decry the concept as such or interpret the computer's impact as destroying, eliminating, or deconstructing it. What is involved is not an attack on it, or a repudiation of its mathematical legitimacy and usefulness, or a refusal of the validity of infinitary mathematics pursued within its own terms, but rather a fresh realization or re-cognition of its deeply ideal status in the light of computational methods. In the light, that is, of an uncompromisingly finite Agent who, as a result of this very finitude is able to sanction use of the concept on the level of the Subject and Person, but as a convenient way of speaking, a permitted abuse of language, that despite it 'literal' meaning has no empirical content. However, it would be naïve to think that such a re-cognition will not impinge ultimately on the status and unchallenged cultural prestige of classical, infinitistic mathematics.

PART II

PART II

Four

PARALLEL SELVES

> I hazard the guess that man will be ultimately known for a mere polity of
> multifarious, incongruous and independent denizens.
> —Robert Louis Stevenson, *The Strange Case of Dr. Jekyll and Mr. Hyde*
> (1886)

Technoid Subjects

Without question, an irrevocable change is happening to the individual self: the thing thought to be fixed and definitional of human identity is becoming unmoored as the technological upheaval transforming the landscape of Western culture makes itself felt deep within our heads, within our subjectivities, our personas, our psyches. Multiple networks of person-to-person connectivity, forms of disseminated and dispersed intelligence, many-leveled mediation—enmeshed in a proliferation of virtual modes of agency and deferred presence—are installing a new psyche at the cultural intersection of these techno-effects and our ancient en-brained bodies.

Raymond Barglow, posing the question of 'self' in terms of the "crisis" engendered by the "information age" asks: "But who is this 'self' . . . and why is it at this particular juncture in the history of Western societies the very identity of the self becomes problematic." (Barglow 1994, 1) But perhaps the first question to ask about the psyche in a technologized milieu is not really one of identity and persona, of 'who' the emergent self is, but *what* and *how* is this self. How is it assembled and transformed by machinic processes, ubiquitous mediation, and ever smarter, more interactive techno-systems? How and by what means do the technological ecologies in which we are immersed alter our consciousness and re-create human psyches?

The understanding that they do is not new. That media and their ecologies restructure consciousness has been widely recognized within many different analyses of the role machinic systems, material practices, and techno-apparatuses play in facilitating new forms of subjectivity and altering human psyches.[1] For Félix Guattari, for example, "Technological machines of information and communication operate at the heart of human subjectivity, not only within its memory and intelligence, but within its sensibility, affects and unconscious fantasms." And, especially relevant here, he singles out what he calls "a-signifying semiological dimensions" of machinic change, effects wrought by technology that escape a "strictly linguistic axiomatics" (1995, 4); effects which bypass or operate beneath language and before signification.

Whatever their ultimate contribution to the signifying and discursive action of a medium, these extra-lingual dimensions of machinic change operate in the first instance through the body of the one who uses and is used by the medium in question. They include practices, routines, and patterns of movement and gestures, as well as proprioceptive, haptic, kinematic, dynamic, and perceptual activities which are either mobilized as part of mediation or are part of the background conditions for the possibility of the medium—activity in the body of the user as a result of changes brought about in order for mediation to take place. Thus, as we've seen for alphabetic writing, these internal changes embrace an entire neurological apparatus brought into being by the alphabet in order for it to function, and with it a plethora of effects from the discretization of utterance into separate 'words' and the analytic illusions of phonemic thought consequent on it, to the linearization of expression and the minutely dedicated forms of attention and perception involved in the act of reading.

We might see these a-signifying dimensions of a technological medium as lying beneath the medium's radar, as part of its unconscious, giving rise to effects not conveyed or represented by it, not within its manifest content and declared purpose but rather as part of the medium's operating necessities, inculcated, implemented, facilitated, and mobilized by it. The logic of these effects is not one of representation but of enactment: they 'mean' not positively in themselves, nor relationally as elements of a signifying code, but through their execution, through the effects that ensue from their having taken place. By engaging thus with the bodies of their users (if only by establishing the physical, neurological, perceptual, and gestural wherewithal to function), by facilitating new behaviors, emphasizing some modes of performance and suppressing others, by engaging users in the repetition or avoidance of certain patterns of action, technological media constitute subjects and reshape psyches in particular, medium-specific, directions.²

Shortly we shall look at two particular media, digital computing and imaging, and find that in each case a central element of the psychic restructuring they accomplish is in the direction of the parallel—toward the plural, the internally multiple, and the distributed—away from the serial, the singular, the monolithic, and the linear. A shift occurs in subjective architecture, in psyches: from the individual and monoidal 'I,' to a collectivized, parallel, and porous mode of self-enunciation.

But first the terms themselves: parallel and serial. Whatever their positive content as separate ideas, they function as a duo. Parallel is bound up with the serial: each opposes and excludes and ultimately engages with the other. So before we proceed, an elaboration of the duo itself.

The Serial/Parallel Duo

A fundamental opposition: given two or more actions (events, behaviors, processes, decisions), they can either occur successively, in serial order one after the other (take one tablet twice a day) or they can happen at the same time, simultaneously, in parallel (take two tablets once a day). The first mode dominates narratives, routines, rituals, algorithms, melodies, and timelines; the second, scenes, episodes, harmonies, contexts, atmospheres, and images. Parallelism foregrounds co-presence, simultaneity, and co-occurrence and is exemplified in collaborating, displaying, and networking, while serialism foregrounds linear order and sequence and occurs in counting, listing, lining up, and telling.

The opposition appears and reappears in many familiar places: music (melody versus harmony), symbolic forms (text versus image), arithmetic (ordinal versus cardinal numbers), film editing (Eisenstein versus intercut montage), electrical circuits (series versus parallel), and, of particular interest here, serial as opposed to parallel computing.

The duo is frequently the site of larger cultural and technical battles. Thus, consider the ancient competition between the verbal and the pictorial as the superior means to truth telling; a standoff that in Western culture has deep iconophobic roots in the biblical interdiction of graven images and in the Platonic distrust of images for being, as simulacra of simulacra, doubly untrustworthy. For William Mitchell the battle is ongoing and fought on a wide terrain: "The dialectic of word and image seems to be a constant in the fabric of signs that a culture weaves around itself. What varies is the precise nature of the weave, the relation of warp to woof. The history of culture is in part the story of a protracted struggle for dominance between pictorial and linguistic signs, each claiming for itself certain proprietary rights on a 'nature' to which only it has access." The dialectic is not a Hegelian one; there is no healing of the rift between the two through some overarching unification. Rather, the struggle between word and image has to be seen as carrying the "fundamental contradictions of our culture into the heart of theoretical discourse itself." (1984, 529)

This conclusion is in tune with Susan Langer's foregrounding of the opposition in the form of discursive versus presentational communication. The discursive, typified by language and the use of numbers, shares the character of words in having "a linear, discrete, successive order, . . . strung together like beads in a rosary; beyond the very limited meanings of inflections . . . we cannot talk in simultaneous bunches of names." (1951, 76) Counterpoised to this is the presentational mode, typified by pictures, which are precisely not discursive: "They do not present their constituents successively, but simultaneously, so the relations determining a visual structure are grasped in one act of vision." (86) It is this capacity to be able to cognize an idea, say, with internal parts, one which has relations inside relations, which "cannot be 'projected' into discursive form" that is peculiar to presentational communication.[3]

Note here the inescapable reflexivity in any account of the two poles and their inevitable intermingling. What gets told or shown—and the issue is already there in that alternative—will necessarily be organized serially as a narrative and parallel as a presentation. The result is a reciprocity of status and action between the two modes, which, as we shall see, makes for a widespread and fecund creative principle whereby the poles operate together, impinging on each other as a combinatorial tool to produce new entities across various humanistic, artistic, mathematical, technoscientific, linguistic, and epistemological practices, as well as being a hard-wired binary alternative within biological systems.

In his account of the development of human mentality, Merlin Donald describes memory and knowledge storage as the fundamental agents of change and structure. Thus, when he characterizes the highest form of prelinguistic mental achievement as that of apes—"unreflective, concrete, and situational"—their lives "lived entirely in the present, as a series of concrete episodes" (1991, 149), he invokes a long-accepted binary within cognitive psychology: contrasting the episodic memory of such lives with the more archaic form of memory, the procedural, that preserves sequences of actions.

In terms of human semiosis, episodes and procedures correspond to the opposition between the parallel co-occurrence of the information in a scene and the sequential delivery of speech. The two forms, found in birds as well as all mammals, employ entirely different neural mechanisms, are morphologically distinct and functionally incompatible: "Whereas procedural memories generalize across situations and events, episodic memory stores specific details of situations and life events" (Donald 1991, 151). However, the opposition here does not exhaust the field of memory: with the advent of language a third, conceptual form of memory emerged. But while this adjoined, reorganized and in much of culture dominates the more primitive episodic and procedural couple, it in no sense obliterated it but rather allowed a new, linguistically mediated version of it to operate.

Mathematics is an entire subject organized around and predicated on the serial/parallel opposition. As Tobias Dantzig, in his discussion of the two conceptual moves needed to handle whole numbers observes: "Correspondence and succession, the two principles which permeate all mathematics—nay, all realms of exact thought—are woven into the fabric of our number system." (1930/1985, 9) The first refers to the one-to-one correspondence whereby the elements of one collection are matched or tallied with those of another; the second refers to the process of ordering the elements into a sequence as part of counting them. Thus, correspondence allows one to judge which of two collections has more elements in the absence of any need (or ability) to count them; succession determines how many elements are in a collection.

Number is a serial/parallel construction. But, as Dantzig notes, the opposition is implicated throughout mathematics and beyond. Certainly, serial (succession) as against parallel (correspondence), in the form of dependence of one thing on a given other versus independence of two co-occurrent things, is fundamental to the construction of post-Renaissance mathematics from the coordinate geometrization of the plane to algebraic variables and the notion of a function; it institutes the separation of independent and dependent events and hence the idea of a random variable in the theory of probability.[4]

A different kind of scientific example comes from quantum physics. There is the well-known parallelist phenomenon of superposition in the standard (Copenhagen) interpretation, where all the mutually contradictory states of a quantum system, ghost tendencies that Werner Heisenberg called *potentia*, are simultaneously present but unrealized. This is opposed to actual or 'real' states of the system resulting from measurements (the so-called collapse of the wave function), which occur one after the other. What can legitimately collapse the parallel into a serial; what, in other words, constitutes a measurement—the so-called 'measurement problem'—is a major mystery for such a view. Interestingly, the main rival model of quantum events, the many-worlds interpretation, eschews superposed parallel tendencies and so eliminates the measurement problem. But by positing one totally determined, un-ghostly state at a time in each of a multitude of 'simultaneously occurring' worlds, it replaces parallel unreal occurrences in one world with real occurrences in parallel unreal worlds.

Not only, as Dantzig observes, are the two principles "woven into the fabric of our number system," but they "permeate . . . all exact thought," and, one can add, also other conceptual domains, well outside the field of mathematics or so-called exact thought, to form a ubiquitous and formidable constitutive principle. Thus, in language, for example, "The concurrence of simultaneous entities and the concatenation of successive entities are the two ways speakers combine linguistic constituents" (Jakobson and Halle 1971, 73), a mode of combination functioning at all levels of speech: phoneme as simultaneous bundle of distinctive features, syllable as succession of phonemes, the parallelism of intonation and the words it accompanies, the combined linearity and simultaneous unity of utterance itself, and so on. This principled amalgamation of serial and parallel modes gives rise to an enormous idea machine, a combinatorial technology that permits the signifying, patterning, imagining, constructing, and discovering of an apparently endless plenitude of entities, abstract objects whose viability and creative potential arise precisely from the way they neutralize the very difference between serial and parallel that allowed them to be brought into existence and juxtaposed in the first place.

By way of elaboration, I present three examples: the code of Western classical music, the language of traditional arithmetic, and the mathematical theory of infinite sets. In each case the objects produced—musical compositions, integers, infinite numbers—are manufactured according to a principle of interchangeability which ensures that—despite the evident opposition between them upon which music, arithmetic, and set theory are founded—any parallel object is equivalent to a serial one and vice versa.

In music the parallel/serial opposition is that between harmony, the production of chords, constituting so-called "vertical music," and the "horizontal music" of melody and rhythm (Oxford Dictionary of Music). In classical music, with its enormously rich, intensely specialized mass of composition based on the principle of key harmonies, this folding of serial and parallel into each other is correspondingly complex and detailed. At bottom, however, it amounts to a vast algebra of musical objects combining the two: compositions which arise from the different ways musicians have formulated re-writing and arranging sequential progressions into simultaneous chords and, inversely, of spilling harmonies over time to be the successive notes of arpeggios and the like.

In traditional arithmetic the principle of ordinal and cardinal interchangeability is so ingrained, and the proliferation of objects combining them so effortless and 'natural,' that it is difficult to detach the principle from the familiar idea of 'whole number.' Thus, any ordinal is obviously a cardinal, since one can forget the counting and treat it as an unordered magnitude. The reverse is perhaps less immediate: it implies that any collection, however named or described or defined as a cardinal magnitude can always be 'counted' into a sequence as an ordinal.

In the theory of infinite sets, where one has infinite ordinals, this last characteristic is problematic. Certainly, ordinals, defined to be sets, automatically possess a cardinal magnitude. The reverse however is precisely the content of the notorious axiom of choice, the axiomatic principle needed to guarantee that all sets can be well ordered, that is, counted. Unlike the finite case, there is no guarantee of being able to order the elements of any nameable magnitude. It is no exaggeration to point out that the possibility of this cardinal and ordinal interchange, as posited by the axiom, is the enabling armature of Cantor's infinite arithmetic. Certainly without it neither that arithmetic nor the theory of sets as developed during the twentieth century would have been possible.

If one looks more closely at the workings of the machine interchanging the serial and parallel modes, one can detect what might be called a horizon effect. In each case the technology of production, the means of creating the plenitude of objects—musical compositions, numbers, linguistic items, and so on—is subject to some form of internal limitation, such as a principle of exhaustion or a law of diminishing returns or an unanswerable question, whereby a horizon of the machine is revealed.

For Western classical music composition the system of vertical–horizontal equivalences collapsed early in the twentieth century, when the key-based harmonies which controlled the chord/arpeggio trade-off were repudiated by a movement appropriately calling itself serialist. For set theory the horizon of the machine was revealed through the proof in 1963 of the independence of the continuum hypothesis, which left unsolvable and essentially unresolvable the question of the magnitude of the continuum (as well as the independence of the axiom of choice that allowed the question of this magnitude to be posed). For the classical integers and their arithmetic, the horizon—less obvious and more speculative—arises from the challenge to the orthodox account of the endlessness of the so-called 'natural' numbers, in one form of which the separation between ordinals and cardinals manifests in the infinite case reappears.[5]

These examples, from cognitive ethology, Western classical music, mathematics, human language, and quantum physics exhibit the parallel/serial duo as a creative and organizing principle across many terrains. Furthermore, because of the dynamics it gives rise to and the mutual enfolding of the two principles within specific cultural practices, changes of status, scope, attributed importance, aesthetic worth, and semiotic transparence in one principle cannot but be accompanied by alterations of corresponding depth and so on in the other. The changes we shall chart, the two co-occurrent, synergistic moves to parallel and distributed computing and digital imaging practices, are certainly central to the contemporary technological scene, and the effects of this emergent parallelism are being felt at every level from how we read, write, and see to the ways we understand ourselves as 'selves.'

Computing Subjects

> Computer architectures are ultimately abstractions of how we think about reality.
> —Norman Margolus

Parallel computational processes divide tasks, data, resources, algorithms, instructions, and memory and distribute them among separate but interconnected elements which perform their operations simultaneously. The shift by computer engineers, scientists, and roboticists, from computation conceived and implemented as one move at a time in sequence to parallel processing and the distributed functionality inherent in it, has changed what it means to compute and reshaped the contemporary technological environment.

The separate, distributed computing elements can vary greatly. They can be fully autonomous computers wired together in local area networks or in networks of networks—such as the Internet—that distribute activities across the planet; they can be the stripped-down, central processing units of computers hard-wired to each other to form ultrafast supercomputers; they can be robotic mechanisms, rudimentary finite state machines, whose behavior is governed by a list of simple, preset instructions. The last, when reproduced in virtual form, as cells in an array within the memory and workspace of a single computer, have proved to be a surprisingly powerful investigative scientific tool. The arrays, known as cellular automata, allow a huge variety of parallel processes to be simulated, so much so that simulation—essentially digital modeling—is, as we have seen, a new, third, mode of scientific research, alongside the established ones of theoretical and experimental investigation.

The well-known explanation of starling flocking comes from simulation by cellular automata: the in-flight behavior of each starling is modeled by an individual cell whose behavior is governed by a local rule: keep a fixed distance from neighboring cells. The resulting pattern of movement by the array reproduces in digital form that of the flock. Similar examples include the generation of artificial life forms in virtual habitats and ecosystems which simulate evolutionary possibilities open to them, genetic algorithms that evolve, refining their ability to solve a problem through the feedback of the results from repeated trials, and pattern recognition and learning behavior within expert systems modeled as arrays. A quite different example comes from fluid dynamics. The Navier-Stokes equation in that subject, a major triumph of nineteenth-century partial differential calculus, captures the behavior of an incompressible fluid; its behavior turns out to be simulatable by a hexagonal grid, each cell of which models a single drop of fluid whose flow in and out of it along the six directions is governed by a simple local rule.

The idea of computing in parallel is natural, obvious, and immediate as soon as one looks at the myriad forms of simultaneous action and collective cognition that surround us. This was recognized at the beginning of parallel computation's rise to prominence by computer engineers.

It is clear that distribution of processing or computation is an intrinsic characteristic of most natural phenomena. . . . Social organizations from honeybee colonies to a modern corporation, from bureaucracies to medical communities, from committees to representative democracies are living examples of distributed information processing embodying a variety of strategies of decomposition and coordination. Computation in biological brains, especially in their sensory processors such as vision systems, displays a high degree of distribution. There is substantial evidence that higher cortical functions are also computed (and controlled) in the brain in an essentially distributed mode. (Chandrasekaran 1981, 1)

To this could be added many other examples of collective phenomena such as the behavior of crowds, workgroups, packs, networks, couples, families, and theatre audiences; not to mention nonhuman collectivities from slime molds, reefs, and colonies to every kind of flock and swarm. In relation to the nature of thought, then, computer science's conversion to parallelism over the past decades amounts to a belated recognition of the presence of collectivities at sites long, deeply, and mistakenly held to be the province of individual, serially thinking subjects.

A noteworthy fact about the introduction of computers in the period 1930–50 was how easily and naturally human 'computers'—which is what people who performed calculations were called—were "annihilated by their electronic counterparts." It's as if we had been waiting for these linear devices all along. And so, according to James Bailey, we had. We built calculating machines that "inherited all the sequential ways of expressing and formulating science that had developed over twenty-five hundred years"[6] (1992, 67), machines that perfectly matched a one-step-at-a-time picture of human computers; and, one can add, perfectly dovetailed with the one-thought-at-a-time picture of our interiority delivered to us through conscious, sequentially narrativized, conscious acts of introspection. Not only did we automatically and unconsciously model 'computer' on a notion of an individual human calculator or thinker, but we also likewise structured scientific research accordingly.[7]

The machine that we call 'the computer' is based on an abstract model laid out by Alan Turing. Known as the Turing machine, and providing the formal, mathematical definition of 'a computation' adopted by computer science, it is a dematerialized, highly abstracted and idealized simulacrum of an individual calculating or thinking self. The computations it sanctions are linear, sequential, and one-dimensional.

It's said that Turing arrived at his celebrated idea through introspection, observing his inner self mentally carrying out the individual steps of a proof or a calculation. In any event, the conception Turing arrived at—a linearly proceeding, one-thing-after-another, sequence of separate computational moves, each of which was notated—matched what most at the time assumed, and many still do assume, to be the workings of our minds performing a 'chain' of reasoning or reckoning.

This assumption has allowed it to serve, in Edwin Hutchins's description, as "an origin myth of [a] cognitive science" committed to a computational, symbol-processing understanding of human thinking. Such a science rides on a confusion concerning the locus of computation: "When the symbols are in the environment of the human, and the human is manipulating the symbols, the cognitive properties of the human are not the same as the properties of the system that is made up of the human in interaction with these symbols. The properties of the human in interaction with the symbols produce some kind of computation. But that does not mean that that computation is happening inside the person's head." (1995, 361) The computational architecture fostered by this misapprehension is thus "not a model of individual cognition. It is a model of the operation of a socio-cultural system from which the human actor has been removed." (363)

To elaborate: the conception of human intelligence and thought embedded in contemporary developmental psychology, artificial intelligence, and the cognitive science Hutchins is critical of—'cognitivism'—is individualistic: it understands thinking, primarily and exclusively, as something taking place inside the enclosed mind/brain of an isolated individual thinker. In order to maintain this perspective it has to assign to 'context' all else relating to the cognitive scene. According to such a methodology the context plays no substantive and certainly no constitutive role in the thinking process. This means that cognitivism necessarily considers as marginal to thought everything outside the individual, symbol-processing brain, from the materio-cognitive means of thinking (writing and other technologies) to the socio-cultural networks and collective environments in which individuals habitually operate. But cognitive ethnographical research presents a different picture, one that insists that what is marginalized into a vaguely defined and all embracing 'context' is, on the contrary, a crucial element in *how* humans think, and that not only is thinking always social, culturally situated, and technologically mediated, but that only by being these things can it happen in the first place. This is why it is vitally necessary according to Hutchins to distinguish between "cognitive properties of the socio-cultural system and the cognitive properties of a person who is manipulating the elements of that system"(362), if one is to understand the relations between individual and collective forms of intelligence.

And, to return to our topic, it is precisely these relations between internal self and external other that parallel computing puts into flux, since it is a machinic implementation, not of individual linear thinking but of distributed bio-social phenomena, of collective thought processes and enunciations that cannot be articulated on the level of an isolated, individual self. Its effects are to introduce into thought, into the self, into the 'I' that it facilitates, parallelist behavior, knowledge, and agency that complicate and ultimately dissolve the idea of a monoidal self. Where the technology of alphabetic writing works to construct a 'lettered self,' a privately enclosed mind, serially structured by the linear protocols reading and writing demand, the apparatuses of parallel computing work in the opposite direction.

Whether through cell phones interchanging private and public spaces; through the plurally fractured linearity of so-called multi-tasking; through the manipulation of external avatars of the self in communally played computer games; through engaging in the multifarious distributions of agency, intelligence, and presence that immersion in networked circuits put into play; or through a still unfolding capacity to be in virtual contact anywhere, at any time, with unknown human or machinic forms of agency—these computational affordances make the who, the what, and the how of the parallelist self radically different from its alphabetic predecessor. And they do so in surprising ways, since the effects and disruptions inherent in parallel and distributed thinking are not easily predictable. This is because collective cognition and collaborative thought is open-ended, heterogeneous, un-schematized, and emergently surprising compared to the more transparent and predictable cause and effect logic of linear thought.

It is necessary to observe, however, that this technological porting of parallelism into thought and selfhood encounters a parallelism already present, long before any engagement with machine computation, consisting of many layers of simultaneous activity of the body from the cellular level to the organization of the central nervous system. A parallelist psyche, then, will be as much an intensification of these existing parallelisms as it is a computational planting into a self that knows nothing of such things. I shall come back to this emerging para-self in the next chapter. But I turn now to another medium, visual rather than computational, which induces its own kind of parallelism into selves.

> While marking the closure of the western metaphysical tradition, decon-
> struction also signals the opening of the post-print culture. Deconstruction
> remains bound to and by the world of print that it nonetheless calls into
> question. What comes after deconstruction? Imagology.
> —Mark Taylor and Esa Saarinen, *Imagologies* (1994)

There is a moment in *Notre-Dame de Paris* when Victor Hugo, drama-
tizing the textual displacement of visual images and architecture as the
means of educating an illiterate population, has a priest hold up a printed
book against the cathedral and declaim "this will overcome that." There is
more than one moment in their imagological essay when media philosophers Mark Taylor and Esa Saarinen declaim the arrival of the "age of post-
literacy" in which the book, as educational vehicle, repository of doxa, and
disseminator of knowledge, like Hugo's cathedral, faces obsolescence, and
held up, poised to overcome it, like a pre-literate pictogram returning on
a Hegelian spiral, is the digital image.

Leaving deconstruction and its fate aside, there is no question that the
vast upsurge in the production and processing of digital images and their
novel forms and accelerating employment across contemporary semiotic
practices and modes of exchange—cognitive, affective, informational—
constitutes a radical shift, a powerful new twist in the century-old process
of audiovisual colonization of socio-cultural practices, one that in relation
to 'post-literacy' pits an intrinsic parallelism against the text-bound, linearly organized print protocols of Western alphabetic culture.

In what follows I consider only one slice of this parallelist move, namely
that concerning the single, stand-alone image (photo, diagram, painting,
map, etc.), and within this I focus for the most part on what one might call
the pragmatic—instrumental or informatic—image. This is not to say that
non-pragmatic art images or the forms of the moving image (film, television, video) are irrelevant. Far from it, but they are relevant in different
ways. Put schematically: film is pre-electronic, television is electronic, and
video is digital, so that including even the most cursory of analyses and
comparisons of these media and the effects allied to them would take us
too far outside the theme here of contemporary parallelism and its anti-
alphabetic consequences.[8]

Certainly, the logic of the alphabet is deeply and inescapably serial.

Its laws are those of sequence, succession, concatenation, juxtaposition, and ordering along a one-dimensional line: from the original fixing of the letters in the abecedarium, to sequences of them to form words, sequences of words through linear syntax to form sentences, successions of sentences to form texts, and so on. Alphabetic ordering enables, among other things, dictionaries, thesauruses, indexes, Dewey classification of books, the whole apparatus of bibliographical scholarship, encyclopedias, and concordances, not to mention nonsense verse, crossword puzzles, secret codes, and Borges's dizzying fiction of an infinite self-cataloguing library.

Alphabetic seriality favors the monad, its movement always swerves into a self-contained whole—the isolated, independent, and indivisible unit or organism, the irreducible thing in itself. Not coincidentally, Gottfried Leibniz, who wanted to construct an alphabet of concepts, was the inventor of monadology. This monoid proclivity is rooted in the alphabet itself—a collection of indivisible phonemes that bear no relation to each other and, being meaningless (in contrast to the morphemes that ground Chinese writing), have no links to anything outside themselves, a lack of width and laterality confining their combination and mode of relationality to one-dimensional serial juxtaposition. As the medium in which the legal, bureaucratic, historical, religious, literary, and social business of the West has been conducted, the alphabet's textualization of thought, affect, and metaphysical systems and its shaping of psychic interiority have been so pervasive and all-encompassing as to be invisible. The very concept of 'a person' has been determined by the apparatus of alphabetic writing, communicating, presenting, theorizing, and framing it. "One can always," at least since the Late Middle Ages, Illich and Sanders observe, "avoid picking up a pen, but one cannot avoid being described, identified, certified, and handled—like a text. Even in reaching out to become one's own 'self,' one reaches out for a text." (1988, x) To this one can add numerous other such observations, from Walter Ong's insistence that "writing restructures consciousness" (1982, 78) to Roy Harris's recognition of the medium's facilitating the self's temporality, in that the writing of diaries allowed "integrating knowledge of an earlier self with knowledge of a later self [to become] a semiological process, subject to conscious control and evaluation."(1995, 44)

But mediological recognitions of writing's formation of the self come (and perhaps could only come) at the close of its influence. We are now witnessing a displacement of the written text's hold on the self; a post-literate self is emerging, patterned not on the word—stable, integral, fixed, discrete, enclosing a unique, interior meaning, ordered, sequential—but on the fluid and unordered multiplicities of the visual image. Early on in this emergence artist Helen Chadwick, noting that "the solitary repressive ego harnessed to language is sovereign," goes on to ask, "What if dangerous fluids were to spill out, displacing logic, refuting a coherent narrative, into a landscape on the brink of I?" The consequences for the self's containment, as Chadwick avers—"Before I was bounded, now I've begun to leak . . ."—are likely to be dramatic and intense in their subversion of a self-contained ego. (1989, 29).

Already, before Chadwick's recognition of the Ego's fluidity, Jean Baudrillard proclaimed the "liquifaction of all referentials" (1983, 4) at precisely the point when data started to 'flow' in streams, and alphanumeric solidity appeared to slide on and off the surface of the liquid-crystal screen. Since then have occurred "liquid television," "liquid architecture," "liquid software," "liquid typography," and many other meltings, image streams, and flows. The ongoing vehicle for the liquifaction of the 'I' conjectured by Chadwick is likely to be primarily technological rather than art practices (notwithstanding its multiple representations in her own paintings) in the form of the informatic or pragmatic image, by which I mean information-bearing, instructional, explicatory, and otherwise instrumentally oriented images—maps, diagrams, tables, charts, graphs, schematics, figures, wire-frame renderings, simulations, models, arrays, scans, and so on. We are in the midst of an explosion of such visual artifacts, whose sheer scope, richness of content, heterogeneous means, stylization, subtlety, and communicational depth prompted art theorist James Elkins to provoke the art historical community with the thesis that such non-art images "engage the central issues . . . and are [as] fully expressive and capable of great and nuanced a range of meaning as any work of fine art." (1995, 553) Elkins's provocation is worth pursuing, and one can go farther along the route it opens up: just as alphabetic writing midwifed a certain self, so pragmatic images, more systematically and with more economic and cultural impact than those of fine art, are now capturing the sites where selves are narrativized, shaped, and reproduced and are now opening up itineraries along which new forms of visual subjectivity are emerging.

There are precursors: the pragmatic image has risen to semiotic power twice before, riding each time on the inadequacy of the written word, whose linear logic, sharp boundaries, and strict government by syntax seem too rigid, precise, foregone in meaning, to cope with a wave of novel, rapidly changing knowledge. First, in the mid-fifteenth century, as a result of perspective drawing which produced, according to historian Samuel Y. Edgerton, "a pictorial language that . . . could communicate more information, more quickly, and to a potentially wider audience than any verbal language in human history." (1980, 189). Second, in the latter half of the nineteenth century, in the wake of the extraordinary image fest unleashed by photography and numerous image producing and visual recording instruments, which led Jean-Louis Comolli to speak of a "frenzy of the visible" (1980, 122), a phenomenon inseparable from the coming into being of new forms of imaged and imaging self.

A century or so later a third wave arrived which was immediately perceived as marking a significant disruption: "The means of production of . . . visual imagery is undergoing a mutation as significant as the invention of photography." (Willis 1990, 197) The mutation is inherent in the now ubiquitous and familiar ability, enabled by digital software, to endlessly manipulate, transform, and display an image. One result—post-photography—indexes an optical regime that breaks with that of the photograph, both destroying the traditional verisimilitude of photographs and fostering a multi-image optic in place of a solitary, unitarily conceived image at the center of itself.[9]

According to Walter Benjamin, a unique work of art emanates an aura, an ineffable presence which is lost in a mechanically reproduced form of it such as a photograph. It would seem that digital imaging repeats a similar dynamic of loss on the medium of photography itself, replacing the chemical fixing of light, its presence at the scene depicted and the source of the photograph's authenticity, its immediacy as a direct trace of the real—which is *its* aura—by the unlimited mediation and mutability of post-photography. As a result, the chemical photograph becomes available to a post-photographic aesthetic as a style effect, a way of citing the pre-digital real. And the camera's claim to verisimilitude is displaced onto a 'truth of a sort' representation, onto a choice between more or less 'objective' digitized 'facts,' instituting what William Mitchell has called "an ontological aneurism . . . in the barrier separating fact and fancy." (1994, 189) Of course, 'truth of a sort' is what painting had always delivered before photography upstaged it, so it is perhaps appropriate that one consequence of "the exchange between the different arts . . . in the digital realm" is that photography "has become more painterly." (Murphie and Potts 2003, 78)

A more abstract link between digitality and the era before photographs, a connection between post-photographic subjects and the prephotographic self, can be derived from a remark on the latter by Jonathan Crary. According to this the notion of self—he says "visual observer"—inculcated by perspective and the camera obscura, a "privatized isolated subject," a "unitary consciousness detached from . . . the exterior" with its "metaphysic of interiority" (1988: 33), collapsed in the early nineteenth century under the onslaught of a physiological investigation of sight and a consequent understanding of vision, as physical, material, embodied. This collapse was masked, however, by the realistic claims made for photography and film as 'objective' and truth-producing representational apparatuses. Adherence to the mantra 'the camera doesn't lie' served to cloak the re-institution of the camera obscura's model of the subject producing what Crary calls a "mirage" of this overthrown self. In this light the post-photographic digital image dissolves the mirage and puts in its place a dispersed, material optic which finally jettisons the Renaissance perspectivalism still clinging to the photographic image and its viewing self.

It can achieve this dissolution because such a digitally enabled post-perspectival self results from a process that reaches down to the level of the image's atomic structure. As an optical regime, perspectivalism constructs, and is constructed upon an en-souled individual, a self with a point of view that is a transcendentally specified location mirroring the vanishing point. But digitization, substituting pixels for points, replaces the psychic architecture and "metaphysic of interiority" of the Renaissance individual by an architecture that, because it must be specified in relation to the physiologically meaningful substrate of the pixelated image, cannot transcend the space it physically occupies, and so cannot enact a metaphysical drama of viewing the world from a position outside it. In Crary's terms, the invisible seeing soul has been finally restored to the physiologically sighted body.[10]

The freedom from chemical control enjoyed by imaging software, its manipulation of code rather than imprints, has destabilized the idea of a 'single' image by allowing an image to be constituted in an endlessly varied format: an image can overlap another, be added to, combined, composited with, juxtaposed, superimposed, interpenetrated, and merged with any other image to form just another image. Once, one would have identified these constituted multi-image assemblages as montages or collages, differentiating them from the standard unitarily conceived image of something presented as a single thing at the center of itself. Now such objects, *imaged images*, have become a default contemporary visual paradigm which, by presenting many images simultaneously within a single optical act, calls for a visual self engaged in a mode of parallel rather than serial seeing. The result is a form of visual polyphony with sampled images as voices whose influence extends across contemporary design aesthetics as well as informatic imaging. Along with this are the now ubiquitous devices and apparatuses of visual parallelism which actively displace a linear optic: Graphic User Interfaces that use mouse gestures to navigate multiple screens viewed simultaneously; Picture In Picture television that permits two or more separate image streams to be displayed and controlled simultaneously; videowall arrays of separately programmable screens on which several different but linked narratives can be run and interpreted simultaneously; and so on.[11]

Moreover, the digitized image's ability to be copied endlessly without degradation makes possible the stacked or layered parallelist optic responsible for the GIS (Geographic Information Systems) revolution in mapmaking. As with the post-photography the shift is from chemical traces on paper—the traditional map—to a digitally manipulable, interactively queryable composite representation. GIS composites superimpose many layers of information pertaining to their object onto the same spatial diagram to produce a class of maps emergently different from their paper predecessors. Such maps are no longer inert, static pictures of a world exterior to it, but become dynamic, a proactive tool capable of being changed by and impinging on the world it represents. GIS maps deliver a co-present assemblage of images that had previously to be viewed serially or side by side. Such multi-seeing now dominates cartography from the architectural (building interiors) and urban (distribution of roads, paths, sewer pipes, and telegraph poles) to the geophysical (satellite images of land use, coastline change, troop movements), and the galactic (interstellar dust and star charts).

Like post-photographic images, then, GIS maps evade the fixity of the viewing subject by bypassing the demands of perspectival viewing. They promote a dynamic viewing body, one that 'sees' precisely by refusing the stationary outlook of a single cartographic projection and with it a claim to capture the 'truth' of a single visual scene. Interestingly this refusal of a single viewpoint through the production of composites is beginning to be deployed by photographers and artists in the medium of digital photography as an aesthetic principle in its own right.[12]

But, post-photography aside, within artistic practices such embodied movement can of course be imagined. Thus, David Hockney, speaking of how, in contrast to the requirement of standing in the middle of a circular perspective of Versailles—"my body seemed to have been taken away," "reduced to a tiny immobile point"—was able to revel in the hours spent "wandering up one street, down another, up another" in such a way that "I still had my body," as he viewed the stylized non-perspective of a Chinese scroll depiction of a city. (1993, 128). Or it can be actual movement, the viewing subject moving in relation to what is being viewed, as occurs in video installations. Here, the image is de-framed and becomes mobile, dispersed, distributed in installations that routinely surround the viewer/ participant in space and require—like the perambulatory form of medieval theatre—to be navigated around, physically negotiated, and experienced in place. Viewing becomes subordinated to an active engagement with the installed objects and scenes surrounding a moving eye. "Engulfed by the assemblage of temporal parts, the process of looking" becomes, in Barbara London's words, "as much about the physical experience as the composite memories that live on in the mind." And so such installations "illustrate the dissolution of the seriality of time that characterizes the late twentieth century." (1996, 14)

In terms of the subject this can be seen as the falling away from a one-dimensional, singular consciousness into parallel, distributed co-presence. What is being analogized and narrativized by such installations is thus another version of the self that is emerging in the face of the imagized image. Thus, in addition to the non-art informatic image with respect to which such a self emerges, we have art practices in the form of dynamically or interactively encountered and immersively viewed installations.

. . . On the Brink of 'I'

Once, not so long ago, there was an absolute separation of self and other: an 'I,' identical to itself as an autonomous, indivisible, interior psyche opposed to an external, amorphous collectivity of third persons outside the skin. Now the I/me-unit is dissolving, the one who says or who writes 'I' (a difference, crucial to Western metaphysics, as we shall see in the next chapter, but overshadowed here) is no longer a singular, integrated whole, but multiple: a shifting plurality of distributed I-effects, I-roles, I-functions, and I-presences. Now the 'I' bleeds outwards into the collective, which in turn introjects, insinuates, and internalizes itself within the 'me': what was private and interior is seen as a fold in the public, the historical, the social as outside events enter (and reveal themselves as having always entered) the individual soul in the mode of personal destiny.

Am I, then, plural; are their two or more of me/us? Can I, you, some-one (some 'one'?) have more than a single identity? Be more than one person?

Not a new question, but one posed most recently as a pathology, known as Multiple Personality Disorder. In this, a person with one body (though so differently presented and experienced as to problematize that singularity) can have many identities or "alters" or personas which take psychic control of the body.[13] Though akin to schizophrenia in its popular understanding as 'split personality,' multiple identity is quite different clinically and experientially. Moreover, unlike schizophrenia, a clutch of psychoses whose causes are unclear, Multiple Personality Disorder is acknowledged to be a specific breakdown, a form of dissociation in the normal functioning of psychic integration. Moreover, a majority of those who treat it attribute it to childhood abuse: the abused self separating off into different psychic compartments as a way of achieving a working interior stability. There is no doubt truth in this etiology, though it seems to hanker after a transcendentally unified subject as the condition for the possibility of rational thought whereas, on the contrary, multiples are by no means incapable of reason or indeed of giving a rational appraisal of their own condition.[14]

Moreover, the abuse etiology of multiples has been directly challenged. Thus, Ian Hacking, tracing what he considers to be its iatrogenic origins, situates it within a history of memory, locating it in the "conceptual space for the idea of multiplicity" (1995, 179) constructed by French medicine in the nineteenth century—a space in which patients describe their symptoms which are then looped back through the doctor–patient circuit into confirmatory evidence of a disease. Though sharply argued and historically focused, Hacking's analysis elides the issue of contemporaneity. Even if, as he maintains, sufferers from the syndrome are creating and fitting their symptoms to a pre-given diagnosis, we're still left with the question: Why these symptoms now? Why the irruption of this kind of multiplicity within present-day culture?[15]

If schizophrenics were the "town criers of modern consciousness . . . existing not just as a product of but also a reaction against the prevailing social order," 'sensitives' too easily able to internalize their torn and disordered times (Sass 1992, 372), then perhaps multiples are their successors, more attuned to the multiplex instabilities of twenty-first-century psychic reality. This is not to assert that multiples aren't aberrant and frequently traumatized in their past, but rather—leaving aside their disputed etiology—that their aberrance might serve, at least for now, as "the best paradigm we have for postmodern consciousness" (Shaviro 1995, 1) and beyond that might illuminate a previously obscured but now inescapable multiplicity inside the 'individual' self.

Thus, in at least one neurological account of multiple personality the syndrome is merely a divergence from a normal psychic parallelism:

> The multiple parallel streams of conscious activity which are implied by the multiple personality data, are no different from those which are present in normal individuals. . . . The difference in the multiple personality case is that these processes can be attached to different self representations, and so when re-represented are revealed as the thoughts of different individuals. When only one self representation is available to self-awareness all conscious process, covert or otherwise, is attributed to a single 'me.'" (Oakley and Eames 1985, 237)

According to this, then, multiples, while undeniably aberrant, are closer to all of us than we have ever, from the perspective of a 'natural' unfractured singularity, imagined.

In fact, internally plural selves, psychic multiplicities seemingly paradoxical to the introspecting monadic self, have nevertheless been thought repeatedly within philosophical and literary psychology. But, along with multiple personality, such models of plural mentation have only been taken seriously by mainstream cognitive and neurological science in the last two decades.

Thus, Marvin Minsky's phrase "society of mind" (1987), an early and still cited metaphor for inside-the-head cognitive parallelism, is in fact a rediscovery of past insistences on psychic plurality. A century before him, the idea was widespread. William James wrote of an inner mental multiplicity; Robert Louis Stevenson asserted: "Man is . . . truly two. I say two, because the state of my own knowledge does not pass beyond that point. Others will . . . outstrip me on the same lines; I hazard the guess that man" (and he means individual humans not mankind) "will be ultimately known for a mere polity of multifarious, incongruous and independent denizens." (1886/1979, 82). While for Nietzsche, "The assumption of a single subject is perhaps unnecessary: perhaps it is just as permissible to assume a multiplicity of subjects whose interaction and struggle is the basis of our thought and our consciousness. . . . My hypothesis: The subject as multiplicity." (1968, 260) More than a century before this, David Hume likened the mind to "a kind of theatre" and argued that "the true idea of the human mind is to consider it as a system of different perceptions or existences, which are linked together . . . and mutually produce, destroy, influence, and modify each other. . . . I cannot compare the soul more properly to anything than to a republic or commonwealth, in which several members are united by the reciprocal ties of government and subordination." (1739–40/1951)

These persistently collectivist figures for the human psyche—society, polity, system of existences, commonwealth, theatre, republic—testify to a recurrent refusal to cognize the self as an undifferentiated and private singularity, insisting on it being akin to a public assemblage of multiple agencies. The historian Hillel Schwartz, commenting on multiple personas and doppelgangers and other doublings that emerged at the ends of the eighteenth and nineteenth centuries, sees the phenomenon as a response to social and economic dislocation, to "societal fears of the loss of generational continuity, most acute at the ends of centuries." (1996, 81) Maybe, though Hume wrote at mid-century, and at the twentieth century's end the effect—whatever its relation to societal fears—occurred within a convergence of neurology, robotics, and post-cognitivist theories of mind onto the notion that even simple mental activities, for example counting, are synthetic assemblages of heterogeneous actions, each in themselves not intrinsic to the activity they enable. (Dehaeane 1997, 13)

Though external technological mediation such as computing and imaging inculcate behaviors and induce forms of parallel cognition and perception, it is the preexistence of an internal 'multiplicity of subjects' that is vital to the pluralized selves that emerge. True, the contours that parallelist technologies carve out in the informational landscape create new pathways for the psyche to follow, and hence originate new self-effects, but they do so in the presence of the inner multiplicities that Hume and others identified. Their mediation is that of an external scaffold, a way of allowing these multiplicities to be recognized, to become choate, to crystallize into social practices, and cohere into experientially real, stable, and iterable forms of psychic activity.

There is however ultimately no separation between interior and exterior: inner experiential 'I' and outer collective 'they' fold into each other. All thought, even the most private and enclosed, is from outside itself, socially existing, being publicly mobilized, using and being used by the media and technological apparatuses that surround us, constitutes our psyches—a phenomenon whose articulation demands an ecology of the self and psychic agency that foregrounds the contemporary technologies of the virtual. It is in this sense that one can agree with Merlin Donald's contention that the key principle of the biological and social evolution of individual cognition is its symbiosis with cognitive collectivities and external memory systems, a two-way traffic that allows new cultural formations and technologies of parallelism to reconfigure the thought diagrams inside (as we still say) our heads.[16]

It is no secret that 'human nature,' whatever unchanging universal that phrase so confidently referred to only a few decades ago, is melting, running off in new directions. But if we are indeed bio-techno-cultural hybrids, if 'nature' is now inseparable from social artifice, and the 'human' is an ongoing, open-ended project of mediated self-construction with shifting boundaries and no identifiable *telos*, then these directions are not arbitrary or unconditioned. Nor do they promise, as many fantasize, a post-human future without death in which digital technology will download minds and at last release psyches from their bodies to float and commune in a sea of disembodied information.[17] Indeed, as we've seen, far from escaping embodiment, the action of technology—precisely in the implicit and unconscious effects of its never absent materiality—is never separable from the bodies of its users, which immediately makes any moving beyond, any attempted abandonment of corporeality, incoherent. From the perspective here the antibody to the illusions of the post-human is the recognition of the 'para-human,' since the condition in question is one of horizontal movement, not upwards or forwards but sideways; not linear or sequential but dispersive and parallel; not going beyond but an expansion, a multiplication, and intensification of what was always there; a new realization of the past and its futures, and with this a recognition of the incipient plurality of a psyche in the process of becoming beside itself.

But what is involved in becoming beside oneself? In experiencing plurality? How does one accede to the para-human? The process is not to be identified with imitating, reproducing, splitting oneself; or identifying with, or assimilating another; or being reborn as a new being (though it can couple with and be traversed by all these). It is rather a form of a-temporal change, becoming party to a condition other than one's own, a question of self-difference, of standing to the side of the single, monadic 'I.' Or, in terms of a meditational practice such as Zazen, of sitting beside oneself to attain a state of mindful un-self. It's easier perhaps to talk in attributes and processes: of an interactively emergent psyche, a self assembled on the neuro-cultural interface that, as we've seen, is both internally and externally collective, a self distributed and in excess of unity.

But one can still ask: What does it mean, where does it go, how will it feel, to be so? Can I, you, those yet to come, really not be what we have (felt to have) been for so long in Western culture, an 'I' that is before all else, as a condition for all else, an enclosed, individual, indivisible, opaque, private, singularly rooted Me? Can the self function in ways other than this organized arborescence? Can it manifest the qualities of what Deleuze and Guattari, in their all too generous vocabulary, call a rhizome? Can it be a network? And do we desire to be so, to become co-present, hetero-geneously connected to ourselves, pulled in different directions from the future, always in the process of becoming multiple and parallel, beside our selves with glee and dissolution, intermittently present to ourselves, each of us a para-self, a psyche on the edge of its seat?

Is it, though, a question of desire and choice? Aren't we, the choosing agents, being dragged, like it or not, into a world of ever more numerous co-happenings? Aren't networks and the relentless co-presencing and distribution of the psyche they facilitate already starting to control the sites where subjects are produced? Aren't parallel computations and multi-images and virtual agencies and a ubiquitous computing and ever smarter and interactive digital environment already deeply embedded in communicational, pragmatic, and art practices and in the heads of the subjects who encounter them? No doubt. And one doesn't have to be a technological determinist to be certain that more will follow. But the question can be more interestingly put in aesthetico-ethical-political terms, in terms of human affect and dreams and visions of our powers, and then the question could not be more real. Or more urgent. It asks whether we wish to embrace the para-human, to deploy the formidable resources of the digital to embrace the other of and within the human, to become different from the isolated monads we were, or rather have for so long thought we had to be? An embrace not made in obeisance to some transcendental authority external to 'the human,' but one that eschews any such, one that seeks, in Antonin Artaud's words, to be done with the judgment of God. Or what amounts to the same thing here, to understand that there is no end of the world and no final judgment, but only humanly created agencies with the magical—ghostly—ability to use machines and technological media in order to look at the human from outside the sphere of human looking.

We can, I believe, embrace the para-human, to begin—haltingly, with confusion, pain, wonder, inevitable resistance, nostalgia, feelings of loss and dread, and moments of intense liberating pleasure, not to say joy and surprise—to become plural 'I's able to be beside ourselves in ways we're only just starting to recognize and feel the need to narrate. There is surely no longer any doubt that something large and unquantifiable is happening across the planet on the outside of our skins and inside our heads. A technologically mediated transformation of the 'human'—global, all encompassing, and seemingly inescapable—is being made by us to happen. Within this upheaval we are going parallel and starting to become beside ourselves, as I've tried in a very schematic way to indicate. We are living through tumultuous, dizzying times on the cusp of a new era; times spanning a seismic jump in the matrix of human culture, which looks to be as momentous, epoch-making, and far reaching in its consequences as the invention of alphabetic writing.

Five

GHOST EFFECTS

> A lady once asked me whether I believed in ghosts and apparitions. I answered with truth and simplicity: No madam! I have seen far too many myself.
> —Samuel Taylor Coleridge, *The Friend* (1809)

> We're living in a supernatural world. . . . We're surrounded by ghosts.
> —Jennifer Egan, *The Keep* (2006)

My concern here is to illuminate the way communicational media can facilitate new psychic entities and objects of belief. Such facilitation occurs when a new medium confronts and absorbs its predecessor. What I have to say centers on the concept of self-reference as this occurs in the enunciation of presence or agency most familiar as, but not confined to, the 'I' of speech, and interweaves two narratives: one about the virtual, the other about ghosts.

The 'I'

But before we begin, let me indicate certain obstacles put in place by the traditional linguistic approach to 'I.' The word 'I' belongs to a family of words linguistic scholars call deictic elements or *indexicals*, on account of their indexing or pointing in some manner to the context of their use, and they characterize them as linguistic expressions whose reference shifts from utterance to utterance, major examples being personal pronouns such as 'I,' 'she,' and 'you' as well as indicators such as 'here,' 'now,' 'today,' 'yesterday,' 'this,' and 'that.' Emile Benveniste, in his classic study of pronouns in Indo-European languages provides a definition of 'I,' the most puzzling of them. "To what does 'I' refer? To something exclusively linguistic: 'I' refers to the act of individual discourse in which it is uttered, and it designates its speaker." (1971, 226)

Benveniste's definition, or more properly the written philosophical and linguistic tradition in which it operates, is problematic with respect to the project here on two counts. The first problem is with the claim, implicit in his definition, that the 'subject' is an exclusively linguistic, incarnate construct, that subjectivity is divorced from and somehow independent of embodied experience and affect: "However subjectivity might be posited in phenomenology or psychology, it is but the emergence into being of a fundamental property of language. He who says ego is 'ego'" (Benveniste 1971, 224). Once the subject's concrete embodiment is restored it becomes questionable to insist that all subjectivity is based on, determined by, originates in, or cannot exist in the absence of language. To do so excludes forms of experience, self-aware sentience, and extra-lingual modes of self-recognition and enunciation that permeate the speaking, 'I'-saying subject, most notably those associated with and delivered by the medium of gesture.[1] Not only gesture in its triple relation to speech, but gesture enshrined in geometrical diagrams, and the many codified and elaborated gestural regimes that engender subjectivities in rhythm, music, song, dance, and a host of devotional, sensual, athletic, aesthetic, and meditational body practices.

The second problem is that the tradition suffers from a certain medium blindness, an unawareness of the effects of itself *as a medium*, an opacity in other words to the work that alphabetic writing cannot but perform on its subject matter. Thus the divorce from the corporeal just noted is already a consequence of the alphabet's handling of speech and can hardly serve as a means of examining the mediological effects of alphabetic writing. In fact Benvenistean linguistics, not to mention the prevailing discourse of the subject (as exclusively a linguistic abstraction) it has influenced, attaches no theoretical importance to the radical discontinuity between textual and vocal enunciation, in particular between the spoken and the written 'I.' This, is despite the fact that Benveniste's definition permits no extension from voice to text, no analogous formulation of itself which would capture the written 'I': not only is there is no unique, identifiable "act of individual discourse" of a textual 'I' to anchor the definition (the putative writer is absent, and an individual act of reading is not an enunciation), but, unlike the one who speaks 'I,' the one who may have written 'I' or (which is not necessarily the same thing) is designated by the term, could be a machine, an actor, or an imagined narrator in a fiction, an agency who might or might not exist, or a self-announcing object. The long and un-troubled conflation of spoken and written 'I's within the tradition in question has masked the possibility that the medium in which the announcing of self-presence occurs might be of importance to its phenomenonological character, its status, to the effects open to it, and to its fate in relation to other media. This being so, a linguistics that fails to separate the spoken from the written 'I,' one which treats *lingua* and the writing of *lingua* as interchangeable—"The so-called science of linguistics has studied writing, not speech" (Powell 2002, 123)—is in no position to elucidate the medium-specific nature of the writing and written subject.

If self-enunciation and the production of subjectivity are indeed medium specific, then it becomes obligatory to frame any account of 'I' accordingly. To this end we can consider framing the project here by considering four media—gesturo-haptic, speech, alphabetic writing, and digital networks—each with its particular mode of self-enunciation in a series involving three re-mediations: from a-linguistic gesture to speech, from speech to its alphabetic inscription, and from alphabetic writing to digital networks.

gesture → speech

According to a recent account of the evolution of language (Deacon 1997), of which more presently, the move from dumb self-gesturing to human speech involved neurological changes over thousands of generations and was moreover responsible for an inherent feeling of ghostliness in the one who speaks 'I.' The move is from an a-linguistic 'I' mediated gesturo-haptically, that is, a proprioceptively continuous 'me,' experienced and projected through self-touching, posture, and visual signaling to the 'I' of spoken language. The gestural medium here is the precursor of the currently observable human visual and communicational gestures which co-evolved with language and has filiations with contemporary gestural forms used by self-recognizing primates, dolphins, and elephants. In human language the gesturo-haptic body is localized within utterance to form the vocal gestures of speech, manifest as prosody or tone of voice joined seamlessly within speech to the words spoken.

speech → writing

The move from the medium of speech to its inscription was a technological innovation that transformed oral societies and, in its alphabetic form, inaugurated Western culture. The remediation in question recalibrated space and time and reconfigured the possibilities of agency and presence. Within the 'speech at a distance' or virtual speech that emerged, vocality was eliminated, and the communicational and affective work of the voice's gestural apparatus was retrieved and redefined through forms of written prosody—mimetically in poetry and transductively through the development of prose styles—at the same time as the body of the speaking 'I' is replaced by an incorporeal, floating agency of the text.

writing → networks

The shift from the medium of alphabetically written texts to their virtualization inside digital networks indexes a contemporary and ongoing upheaval of undetermined scope whose effects surround us. Within it, the fate of the written 'I' is to be subsumed—assimilated into, repositioned, and overlaid—by digital forms of self-reference, by modes of self-enunciation intrinsic to, and only possible within, the instantaneities of a digitally recalibrated space-time and reconfigured agency/presence facilitated by interactive and distributed electronic networks.

The three remediations enact their recalibrations and reconfigurations through the body, at every level from its neurological structure and modes of perception to the shape of its semiotic envelope. The latter is dominated by physical gesture, and we can observe gesturality tracing out a certain itinerary along the chain of remediations: from a dumb a-linguistic base to its vocalization in speech to its eliminative transduction in the text to its reemergence as the means of navigating the infosphere; the latter initiated by the computer mouse and continuing through its gesturo-haptic successors such as motion capture.

In what follows I shall touch briefly on the first remediation and elaborate somewhat on the third, but it is the second, the effect of alphabetic writing, that will figure most. Thus, to return to our starting point, we shall see that the metaphysical consequences of the West's alphabetic literacy in both its Hebrew and Greek forms rests precisely on the elision of the mediational work performed by conflating the written 'I' with its spoken predecessor, an elision crucial to the creation of the West's three founding ghost abstractions: namely, the invisible God, the un-embodied Mind, and the Infinite mathematical agent. Or so I argue.

But before confronting these ghosts I want to introduce the other narrative thread here—the phenomenon of the virtual.

Virtual X

The *virtual* names a powerful, pervasive operation of contemporary network technologies, an operation whereby an activity—seemingly any activity X—of contemporary culture is confronted by and interacts with a virtual form of itself. Until barely three decades ago 'virtual' was an unassuming modifier connoting something insubstantial, imaginary, not quite real—"in essence or effect not in form" (*Webster's Dictionary*), a usage encompassing certain scientific and mathematical instances referring to thought experiment deployments of unreal or non-real entities, such as virtual rays and foci in optics, and virtual forces and work in mechanics. Now, the virtual operates everywhere in the contemporary scene. Its techno-scientific instances have mushroomed into virtual space, virtual particles, virtual waves in quantum physics; virtual machines and virtual memory in computer engineering; virtual organisms and environments in artificial life; virtual molecules in drug design. And beyond these technical senses, virtual shopping, virtual books, virtual universities, virtual bodies, virtual sex, virtual reality, virtual subjects, virtual history, virtual keyboards, virtual warfare, and an expanding slew of virtual entities, processes, and forms of agency permeate social life.

These virtual Xs are plainly the outcome of electronic technologies enabled by digital computation, but one can question whether the phenomenon of virtuality they exhibit is itself electronic, whether computation or communicational networks are intrinsic to the virtual.[2]

Certainly, the vectors of decomposition and reconstitution that inhabit the virtual, whereby processes are wrenched free of their governing temporalities and original milieus to be displaced, recontextualized, and relocated in a virtual elsewhere, are in themselves abstract, generic effects of mediation with no necessary connection to specific technological regimes, electronic or otherwise. These vectors make themselves felt as a series of discontinuities, shifts in the familiar attributes of everyday psychic and social reality. Thus what was always predominantly serial and linear—computations, texts, narratives, and vertical chains of command—are being overtaken by parallel and horizontally structured forms; what were assumed to be individual, private, and isolated—mind, subjectivity, thought—are increasingly public, distributed, and communal; what were always thought to have been interiorly and endogenously formed human processes—the production of affect and desire, psychic development, the 'human' itself—are increasingly revealed to be exogenous and multiply assembled from outside themselves.

Each of these shifts by itself marks the site of a psychically dense, far-reaching cluster of socio-cultural discontinuities. But they are interconnected, and taken together in concert indicate the emergence of a large-scale, radical transformation in the discursive and experiential fabric of Western culture. How large? How radical? How might one frame a transformation of this magnitude? What could be a comparable cultural disruption? Let me offer a provocation.

Virtuality is ancient. Far from being tied to contemporary electronic technologies, its lineage long antedates its current technological matrix. Its present manifestation, the ubiquitous virtual X inundating contemporary life, is the third great wave of the phenomenon. The second wave came with the writing of speech. The first cannot be separated from the advent of human language itself. Further, each of the mediations—spoken language, the technology of speech-writing, contemporary digital and post-digital media—midwifes virtual effects into being at the site of an irruption, a coming apart of a previously self-sufficient and seamless whole. The result in each case is a dissociation which restructures consciousness to produce modes of presence, agency, and self-representation that were unavailable, or, even if available as imagined constructs, were unimplementable in any stable, practical form.

In particular and of principal interest here, contemporary technologies of the virtual are dethroning the habits and protocols associated with their alphabetic predecessors, whose authority and ubiquitous presence have dominated Western culture since its inception. The resulting displacement of concepts, practices, and psychic entities that were maintained by reason of written mediation, transforms the latter into virtual forms of themselves creating milieus and operating within conceptual environments outside their horizons. I shall return to these horizons later, but my emphasis for the most part will be with the accomplishments, that is, the mediological effects, of the technology of alphabetic writing rather than its overcoming, with certain disembodied entities that emerge from written 'speech at a distance,' from virtual speech, and I shall look at three entities—God, Mind, and Infinity—each of which arises as a hypostatized agency, a ghost effect within the matrix of this technology. But before this a remark about the first wave of the virtual, namely the ghost effect which emerges as a deep-rooted and seemingly inevitable accompaniment of human speech.

Bio-linguistic Ghostliness

> A concept for which there is no referent, no evidence, anywhere, any place, any time in the entire sweep of human experience, yet one that is vital in many cultures and perhaps in every culture since the Upper Paleolithic Age. Apparently, there is a powerful impulse within human imagination that flows inevitably to this unwarranted fiction.
>
> —Mark Turner, "The Ghost of Anyone's Father" (2003)

To avoid misunderstanding, let me say at once that by 'ghost' I mean something other than the customary spirit of a deceased person and more abstract than "a social figure" signifying a haunting (Gordon 1997). If indeed we are surrounded by ghosts in a 'supernatural world,' it's because our psyches in this world are permeated by new media. The ghosts that concern me here (which in fact are not normally designated as such) are media effects, invisible, technologically induced agencies that emerge, under appropriate circumstances, as autonomous self-enunciating entities. They are medium-specific, their character and action being intrinsic to the medium in which their existence is manifest, and their efficacy as objects of belief and material consequence derive from their unacknowledgement—their effacement—of this very fact. Ghosts may be particularly postmodern entities (I'll get to that later); they're also ancient: in the history of the West certain disembodied signifiers—ghosts of writing—have given birth, as we shall see, to theological, metaphysical, and mathematical thoughts and beliefs.

Though distinct from specters such as individual *revenants* demanding, like the ghost of Hamlet's father, retribution, revenge, and the righting of wrongs, or from identifiable sites of social repression and invisibilization, media ghosts are nonetheless neither free of the supernatural nor the uncanny affect inherent in their disembodied otherness. In that sense, they too might be said to haunt.

But before we get to such things, what, we might ask, is an 'unwarranted fiction'? A concept without a referent? How do things get to 'be'—to be named and sustain belief in their existence—with no apparent evidence for them? In particular, how is it that language is able to refer to impossible actions and nonexistent entities? The question is not one the science of linguistics is prompted to ask, since (following Saussure's structuralism) it takes the structure of an abstractly conceived system of language (*lange*) as its object and brackets out (as *parole*) as an entirely other matter issues belonging to the socio-ontological deployment of speech, that is, questions about the reality or otherwise of what language names, refers to, and talks about.

Whatever its merits for articulating and investigating the grammar, morphology, and syntax of reference terms, and however augmented by 'pragmatics,' a structuralist framework offers no *linguistic* explanation or insight into how concepts with no referents might occur, and why their genealogy might be significant. By contrast, a recent neurobiological account of the evolution of language by Terrence Deacon, which rejects Saussure's division of langue and parole and its account of reference, does just that. According to Deacon, reference, "the correspondence between words and objects," is "a secondary relationship, subordinate to a web of associative relationships of a quite different sort, which even allows us reference to impossible things." (1997, 70)

Deacon's account of the evolutionary genesis of this web and its role in the formation of reference rests on a lengthy neuro-physiological and bio-cultural narrative too complex to reprise here. An abstract of it goes as follows.

In normal usage 'reference' means *symbolic* reference, the use of a *symbol* for something, a capacity which Deacon understands as emerging out of a double layer of two simpler kinds. First, and the sine qua non of any sensory activity, comes iconic reference: a similarity or, better, the lack of a perceived difference, between a sign and its referent;[3] second, indexical reference: which Deacon understands as a contiguity interpreted to exist between a sign and its referent. The hierarchy is completed by a third kind—symbolic reference—made up of representations of relations between indices. Whereas the first two are widely shared across species, all nervous systems exhibiting some form of iconic and indexical reference (albeit across a huge cognitive range), the last is confined to humans with only a rudimentary form learnable by some higher primates.

The terms icon, index, and symbol are from Peirce's semiotics, and Deacon's deployment of them, though he ranks the triad in a way Peirce never did, adheres to Peirce's axiom that all ideas are essentially transmissions of signs organized by a semiotic logic that is the same for communication processes inside and outside the brain. This means the deployment of icons and indices relies on inferential and predictive powers implicit in their underlying neural mechanisms, which "are not physically represented but only virtually re-presented by producing . . . responses like those that would occur if they were present." (1997, 78) Because it only ever handles representations of events and not events themselves, the brain itself imposes no separation in principle between the hypothetical, the possible, the impossible, and the actual—all are treated as representations of representations.

This allows Deacon to show how symbolic reference can arise from the concrete, biologically constrained here-and-now world of icons and indices via a process of Baldwinian co-evolution of physical and cultural processes in which neurological structure and language ability, brain, and the capacity for speech, interact: changes in the newborn's brain, manifest in its cognitive abilities, exert a selective pressure on which features of language are learnable and which not, which in turn feeds back to influence the brain to change in certain ways, which further impacts the grammatical structures and semantic possibilities of spoken language available to the infant, and so it goes. The outcome, over thousands of generations, is the emergence of human language in which symbolic or virtual reference, the ability to refer to that which is absent or unreal, emerges from and is distributed among a web of evolutionarily older iconic and indexical forms of reference.

The upshot is that human speech, inhabiting language, "the impossible to state state of being-in-language" (Giorgio Agamben) is from the standpoint of evolutionary biology an impossible to separate entanglement of two modes of self-reference, experienced by all talking beings as an ever-present doubling, an unstable alternation between an 'actual,' pre-linguistic indexicality, the dumb haptic and gestural self-pointing that extends proprioception, and a symbolic, 'virtual' indexicality in play as soon as one speaks. Put differently, the virtuality introduced by human speech ruptures the pre-linguistic subjectivity, the gestural *umwelt*, of an individual, and thereby allows an escape from what Merlin Donald calls the "mimetic stage" of cognitive development [1991, 1998] into the symbolic domain of the spoken 'I.'

Thus, the spoken 'I' cannot but fold within itself the unexpungeable invocation of other selves, not only the 'you' to whom utterance is addressed and without whose assumed presence human speech is impossible, but also the generic 'they,' the other, reference to whom is always symbolic. On the viewpoint that human language evolved because we are social agents such a conclusion is inevitable: "One cannot conceive of oneself *as* oneself without also conceiving others as self-directed, egocentric agents."[4] (Corazza 2004, 348) This interfolding of actual and virtual reference, which enters into every occasion of linguistically mediated auto-reference, might be the basis, Deacon suggests, for a biology of ghostliness, a natural origin for the psychological salience and subjective reality of non-mortal, non-natural entities: "The symbolic representation of self," he says, "provides a perspective on that curious human intuition that our minds are somehow independent of our bodies; an intuition . . . translated into beliefs about disembodied spirits and souls that persist beyond death." (1997, 454)

In the presence of the symbol, then, all self-representation loops through the physically absent but virtually present social other. Split off and reified, this virtual component of auto-reference is externalized as a being—a ghost-spirit of the self. In this sense, one can say that speech midwifes the first consciously perceivable out-of-body phenomenon, felt as the uncanny physical experience of a body speaking itself. As such, it might be said to echo the evolutionarily older rupture (shared with mammals), namely the out-of-the-womb experience of birth which delivers the body itself into hominized life. Subsequent manifestations of the virtual, such as that associated with writing, engender other disembodied entities, other forms of remote agency, with their own forms of ghost presence.

Deus ex Machina

Ghosts cling to communicational media. Sometimes a medium's connection to ghosts will be immediately evident. In 1844 Samuel Morse ushered in the digital era by stringing a telegraph wire between Washington and Baltimore, and his first official message was, "What hath God wrought." Within half a dozen years one answer to this question emerged as hundreds of people in New York State and many thousands across America took part in rapping sessions with the dead. Women, functioning as mediums between this world and the next, put yes or no questions to a spirit ascertained to be present. They would get two raps for yes and three for no or a series of raps if the answer was a number or a letter of the alphabet. Thus was spiritualism, a faith which offered empirical—'scientific'—evidence of the immortality of the soul, wrought by the telegraph: those who took part in séances called themselves scientific 'investigators' who could observe 'demonstrations' carried out under 'test conditions'; the new religion named its newspaper *The Spiritual Telegraph*. Over the next decades the telegraph model of spirit communication was augmented by other media and more accomplished mediums who experienced spirits moving their hands to produce automatic writing (typewriter) and by spirits controlling their voices to produce trance speaking (phonograph, telephone).

By providing real-time—live—communication with remote, invisible, and unknown persons, telegraphy allowed spirits of the dead to come to life, crystallizing, in the presence of socio-religious yearnings and an obsession with mortality, a population of ghosts—remote, invisible, disembodied agencies who, replacing electric by psychic power, communicated in real time by (appropriately enough) Morse-code like taps through a human medium. Electrical and animal magnetism have never been more intimately connected and mysteriously interchangeable. Of course this contacting of the dead didn't occur in a vacuum. "America," Erik Davis observes, "already had Shakers channeling Native American heirophants and Stateside mesmerists interrogating spirits through their zonked-out patients." (1998, 60) But it's hard to imagine spiritualist beliefs about the lives of the dead capturing the minds of a wide swath of nineteenth- and early twentieth-century artistic and scientific intelligentsia—from Thomas Edison to Arthur Conan Doyle—without the ever-present telegraphic model of table rapping and its Ouija board elaborations. There was also the phonograph's direct connection to the dead. Thus Edison, impressed by the "startling possibility" of the phonograph preserving the voices of the deceased and convinced of the survival of the human personality after death, attempted until he himself died, to increase the sensitivity of his machine to a point where it could pick up directly, without the need for a human medium, the faint vocal vibrations that emanated from the dead.[5] Edison's project lives on in a contemporary format as research into Electronic Voice Phenomena and Instrumental Transcommunication.[6]

As the telegraph was propagating its ghost communicators, initiating what Erik Davis calls "the electromagnetic unheimlich" (1998, 68), the photographic plate was revealing its own contributions to the uncanny: luminous body images, ghostly apparitions, and ectoplasmic traces of previously invisible and unsuspected paranormal forces. And in the telegraph's wake, each new technology of communication—phonograph, radio, gramophone, telephone, X rays, film, radar, TV, computer networks—added its own spectral effects to the layers of uncanny visitations, hauntings, channelings of the dead, and associated ghostly activities.[7]

But all these modern media technologies and their ghosts ride on the back of the much older primal and enabling technological medium of writing. Telegraphing is alphabetic writing in live time (whose contemporary version is mobile phone texting). Millennia before electric there was the *unheimlich* of the text induced by the psychic shock of speech from elsewhere, disincarnate voices, silent speech miraculously and spookily locked in scrolls of skin and papyrus waiting to be ventriloquized into living speech. Long preceding the telegraphic tapping out of 'I,' the ability to name and announce oneself in writing across distance and time was the source of uncanny messages and hauntings from the grave.[8]

If, as suggested, the very ability to speak 'I' fomented belief in the existence of a symbolic entity, an incorporeal spirit of the self separate from the speaking body, might not the phenomenon be more general? Might an en-ghosting occur when any new communicational medium—which is always a re-mediation—takes hold and is used reflexively? Spoken language is a remediation of the pre-linguistic system of human gestural communication, and the passage from a dumb haptic, self-pointing 'me' to spoken 'I' is a shift from actual to virtual, from a proprioceptive self to its symbolic form. If X is 'me,' then virtual X is spoken 'I,' and the belief in a ghost spirit of the self hypothesized by Deacon arises from experiencing the two self designations against each other. What then of the remediation of speech by writing? Like any medium, writing projects an arbitrary or universal user of itself. This mediological parallel to the indefinite pronoun 'one' is a virtuality, an abstract would-be figure distinct from any potential or actual reader or writer located in space and time. In the writing of 'I' this figure, hypostatized as an agency circumscribing all possible referents of the pronoun, can, under appropriate conditions, become the putative, self-referring originator of the text containing it—an invisible, absent writing agency, detached from the voice, unmoored from any time or place of origination, and necessarily invisible and without physical presence.

At the beginning of the alphabetic West two figures inhabiting this description emerged. Two quasi-human agencies, master ghosts of the alphabet, who/which have since constituted major horizons of Western religious thought and intellectual discourse, came into being. One was God or *Jahweh*, the external monobeing conjured out of their tribal god by the Israelites; the other Mind or *psyche/nous*, the internal organ of thought conceived by the Greeks to exist as a non-somatic—'mental'—agency.

These two agencies could not have occurred in more different social, historical, cultural, and intellectual milieus. Nonetheless, each of the Hebrew and Greek encounters with alphabetic writing gave rise in the sixth century BCE to a supernatural, disembodied agency. They did so (as far as is known) independently of each other through the same mediological move of exploiting and being captured by the ghostly possibilities inherent in the writing of 'I.'

The Jewish 'I': Jahweh

In the religions of the book, the monobeing—invisible, ubiquitously present, immeasurably absent—speaks to its Jewish, Christian, and Islamic adherents through a human intermediary who writes or dictates the divine utterance in a book made holy and sacred by that being's presence 'within' it. My account here is confined to the Judaic version of the monobeing, *Jahweh*, and its book, the *Torah*, and focuses on how the Jewish God came to be—to be written and to exist—inside it.

As an idea it would appear that monotheism did not originate with the Israelites. According to historian Richard Gabriel, "The tendency toward monotheism in Egyptian religion was very old indeed. . . . The Egyptians arrived at the proposition early on that all gods were but manifestations or permitted forms of one god." (2002, 34) This tendency emerged as a full-blown religious practice in the fourteenth century BCE in the hands of the pharaoh Akhenaten. His religion was not based on written revelation and not coincidentally was short lived. Gabriel hypothesizes, however, that though lasting little more than a generation within Egypt its effects were transmitted through the figure identified as Moses to form the basis, a thousand years later, of Jewish monotheism.[9]

Whether this subterranean transmission is historically plausible is a matter of argument, but the particularity of *Jahweh* as a certain kind of being possessing a certain kind of agency and appearing with a certain sort of announcement requires a more complex lineage than the resuscitation of a prior Egyptian idea, if only to explain the fact of the inseparability of His appearance from a particular alphabetic text. It is the combination of the worship of a supreme being tied to His written revelation—God inside the Book—that needs to be explored, a combination by no means 'natural' or obvious. Indeed quite the opposite for Gilles Deleuze and Felix Guattari: "Monotheism and the Book," they remark, "is the strangest cult." (1988, 127) Strange or not, how, we might ask, did alphabetic writing, Jews, and an invisible monobeing get entangled? How did a manmade array of written marks on a scroll of sewn-together animal skins become a 'holy' site, a fetish, for the presence of the eternal invisible God?

The seminal event in the *Torah* is God's encounter with Moses on

Sinai at which his covenant with the Israelites, foundational to the ethno-theological narrative of Judaism and crucial to its transcendental claims and unique religious identity, takes place. It is told in the book of Exodus twice. First, the deity *inscribes* his commandments, "And he gave unto Moses . . . two tables of stone, written with the finger of God" (Exodus 32:16), then a few verses later, he *speaks* them, "And the Lord said unto Moses: write thou these words: for after the tenor of these words I have made a covenant with thee and with Israel." (Exodus 34:27). Which is it? Did God speak the commandments as a living utterance into Moses' ear? Or did He deliver them as an already finished written text? Is the question meaningful?

The Hebrew Bible's self-ascription as an historical narrative of real events suggests that it is. However, though framed as historical truth, presented as a continuous itinerary of the Jewish people founded by Abraham and understood to be authored by Moses, chosen prophet of God's word, the five books of the Pentateuch are nothing of the sort. Two centuries of biblical hermeneutics and more recent epigraphical, historical, and archeological scholarship have shown the Hebrew bible to be a skillfully wrought assemblage of diverse and at times contradictory texts by different hands, in different milieus, with different agendas, dating from the tenth to the sixth centuries BCE, rewritten and altered by priest-scribes during and after the Babylonian sixth century exile.[10] According to this there are two 'Gods,' one from circa 900 BCE, the other from the period of the Torah's priestly redaction some four centuries later. The earlier one is the God of the Israelites, a tribal appropriation of El—the chief Mesopotamian god (hence Isra-el, El-ohim, Gabri-el, etc.)—who promises to smite their enemies and jealously demanded that no gods be worshipped before him. Evidently the earlier God is not yet the one and only *Jahweh*: monolatry has not yet become monotheism. It is the later God, the one who *writes* the covenant, who is *Jahweh*, the monobeing for whom other gods are not merely inferior—"Thou shalt have no other gods before me"—but false, nonexistent. This appropriation of the tribal God by the monobeing was rhetorically consolidated by the priestly insertion into the earlier book of Isaiah of the hysterically reiterated "I am the lord, there is no other . . . there is no god but me." (Isaiah 43–45).

As a political-theological instrument, the Bible, in Regis Debray's phrase, "magnificently fulfilled its role . . . by fabricating an origin in order to invent a destination." (2004, 30) What, one must ask, was the role of writing in this fabrication? How exactly, by what specifically textual means, was the unique God of 'the people of the book' fashioned for them within its alphabetic interior? How did the fact of his writing and being written constitute *Jahweh* as a monobeing for the book's readers?

From its inception until the Middle Ages reading for the most part meant reading aloud, voicing the text, decoding the groups of alphabetic letters by supplying the vocal gestures—the prosody or tone—necessarily omitted from the speech they transcribed, gestures indicating motive and intentions, conveying desire, and engendering affect. To read aloud (or in silent simulation), is to attempt (or to imagine) a retrieval, a serviceable approximation of a tone by relating the text to a social or cultural context of its production, by invoking prior knowledge of the source and origin of its contents, by placing the text in relation to other similar texts, by tracing its historical provenance and the author's attitude to the utterance, and so on. But if the context is mythically unrecoverable, if there is no prior knowledge, if the inscription is unique and unrelated to any other, if the author is singular and has no reality outside the text, then any such retrieval, whether spoken aloud or sub-vocalized in so-called silent reading, will be impossible. Defined entirely within toneless writing, the 'voice' of such an author *is* its inscription. What would be the attributes of such an agency, one who/which 'speaks' in a voice absent of all tone?

Human speech is predicated on a circuit of desiring and reciprocating agencies. Utterance projects—is directed to and is impossible without—an already-there embodied addressee, a like-minded, turn-taking other who listens. Tone, the presence of the body in speech and vehicle of its desire and affect, is the means by which awareness of the presence of this addressee is conveyed. But how could a voice neutral in tone—flat, expressionless, lacking all rhythm, having zero affect—acknowledge or exhibit awareness of the presence of the other? Would it not, on the contrary, evince total indifference to the very existence of a listener? And would not such a neutral voice, one that does not recognize my being let alone desire me, induce an emptiness, an experience of absolute—that is, inhuman—terror.

Vocal gesture, a principal means of hominization, the route along which infants become speaking humans, is also one of the earliest means, through prosodic differentiation, by which humans tell each other apart. If prosody is absent from a voice, then the bearer of such a voice is unknowable in an auditory sense as an individual, as a speaking being among others. "Voice," Italo Calvino says, "means that there is a living person . . . who sends into thin air this voice different from all other voices."

(quoted in Cavarero 2005, 1)[11] Thus it is with the alphabetic God who exists and 'speaks' only as a voiceless writer. Short of invoking a plurality of indistinguishable and interchangeable speakers (like identical atoms or mathematical units), a toneless voice can only invoke a singularity, a one-and-only, self-identical entity comparable to nothing outside itself; a monobeing who is not merely one of a kind, but *is* its kind.[12]

Thus, two of *Jahweh*'s principal features, a presence too terrifying "to behold" and his unique status, "There is no other . . . no God but me," are (quite separate from the theological reasons subsequently adduced) media effects, the inevitable characteristics of a god revealed and knowable only in and through a written alphabetic text.

The question remains: how did God come to have a presence inside the *Torah*? The writing of 'I' would seem to point to an agency—an 'I'-er, the person or thing who or which writes 'I' and exists outside—before—the act of writing. But there is no necessity for such prior authorial existence: unlike its spoken version a written 'I' can be fictional, a possibility not refutable in the case of the *Torah*. On the contrary, might it not be the case that the being who says (i.e., writes) 'I' is no more or less than an autochthon of the alphabetic text itself?

The being first appears to Moses through a sign (a burning bush) and hails him: "Moses, Moses," to which Moses responds: "Here, I am."[13] (3:4) He introduces himself through an obscure, enigmatic formula: "And God said unto Moses I AM THAT I AM." This doubled self-assertion is followed by a strange, contrary to grammar, self-designation. Asked by Moses how he—God—should be named as the author(ity) of the message Moses is to deliver to the errant Israelites, the being answers: "Thus shalt thou say unto the children of Israel, I AM hath sent you." (3:14)

The cryptic "I am that I am" and the name "I am" are not peculiar to English. In French, it's "Je suis celui qui suis," followed by "Je suis"; in Luther's Bible "Ich werde sein, der Ich sein werde," followed by "Ich werd's sein." All three are from the Greek "ego eimi ho on," one of whose alternative English translations is "I am the one who is." The Greek in turn is a translation of the Hebrew-Aramaic "ehyeh aher ehyeh" which, besides "I am that I am," has been rendered, consistent with the view that the verb 'to be' existed in pre-exilic Hebrew and Aramaic only in the future tense, as "I will be that I will be" as well as "I am the existent one, the one who goes on being."

Writing, by rescuing speech from oblivion—the fate of the all that is ephemeral and traceless—allows utterance to live beyond itself, thus inventing the idea of a perpetual, unending future and the reality of an unchanging, interminable covenant. This sense of writing-induced continuation, inherent in God's doubled declaration of existence, is enhanced by a form of oral word magic transposed to writing. The invocation of an object or an agency created by uttering its name is a widespread and ancient ontological principle deeply embedded in the *Torah* from its beginning: "And the Lord said 'Let there be light' and there was light," as well as in Adam's naming of the animals into the world. Subsequently the principle reappears transformed into an axiom of Christian theology: Jesus Christ as the word of God made flesh.

The principle, the naming of an entity preceding and determining the substance and existence of it, subverts the supposed externality and independence of language to that which it describes.[14] Deployed here in a supposedly realist or historical text, the principle allows the writing of an event to bring it into existence under the guise of a descriptive report. Thus, not only does the writing of 'I' assume an 'I'-er who must have preceded it, but the existential claim "I am that I am" (equally "I will be that I will be") and the initial fragment of it, the name "I am" ("I will be") become circularly related: each precedes the other. The effect is that of a recursion which, by using its output as a new input, is able to either define or create something in terms of itself. A contemporary example of this manner of recursivity occurs in computing where the name of an operating system, GNU, is an acronym for 'GNU Not Unix,' a name which thereby precedes and enters into the defined meaning of itself. Likewise, "I am that I am" enters and is determined by the meaning of "I am." Interestingly, recursive operations within a computer program are described by software engineers as the program "calling itself" — a locution which captures well the autocthonic effect of the naming formula operating within the written 'I' here.

Jahweh's self-birth from within alphabetic writing at the site of the written 'I' left in its wake an intense and lasting alphabetic fetishism within Jewish mystical and philosophical thought. On the former, one has the written scroll of the *Torah* as a sacred fetish object, subject to ritual, revered as if animated by a spirit, kept in an inner sanctum of the holy of holies, guarded originally by winged warriors, and carried ceremoniously into prayer; an originary relation of God to His alphabetic text transformed within Kabalistic writings into a narrative of God's employment of the letters to make (write) the world into being.[15] On the latter, a more rationalist rabbinical strand celebrated the alphabetic principle itself over its function of writing down the sound of speech. Thus Daniel Heller-Roazen relates how Talmudic scholars, focusing on the delivery of the commandments at Sinai, argued that only the two phrases following the initial 'I' (*anochai*) of the first commandment, "I am the Lord thy God," were audible to the Israelites at the foot of the mountain. A millennium later, Moses Maimonides reduced this further: only a sound, "a strong voice" and not any intelligible words, was perceived by anybody at Sinai other than Moses. Further still, Hassidic scholars in the eighteenth century replaced this sound by the single letter, aleph, which begins the word *anochai*. But aleph, essentially a glottal stop, is for Heller-Roazen strictly speaking no longer a letter representing a sound, but the "blankness of an absolute beginning . . . a mark," not of speech, but "of silence at the inception of speech." (2002, 103) Observe that this contracting series of representations of God's word, each replaced by an initial section of itself, mimics the acrophonic principle whereby the alphabet itself is held to have arisen. One can therefore see in this converging onto a silent aleph, a long drawn-out alphabetic distancing from the El-component of *Jahweh*, a dethroning by the God who wrote 'I,' of the older, speaking God. By the twentieth century the voice of God had disappeared entirely, only writing remains: the triumph of the letter finding its contemporary terminus in grammatology, "one of the postmodern branches," according to Vassilis Lambropoulos, "of the Science of Judaism" (1993, 260); a triumph enshrined in Jacques Derrida's voice-silencing and body-annihilating grammatological mantra "There is no outside to the text."

Finally, there is the question of the visual in relation to the alphabet. Of the religions of the book, Judaism and Islam have been faithful to the second commandment interdicting all picturing of God or Allah. And despite Christianity's obvious abandonment of the commandment made necessary in the face of a physical Christ, the tradition of Western philosophical and theological discourse has been remarkably faithful to it, a fidelity manifest in a refusal, for the most part implicit and unexamined, to countenance images or other forms of visual semiosis within its purely alphabetic texts. Unlike other inscriptional systems such as Egyptian hieroglyphics or Chinese characters or Mayan glyphs or scientific and mathematical symbols and diagrams, alphabetic writing lacks any visual connection to what it represents. As a consequence, the tradition has failed to raise the question of whether *Jahweh*'s (theologically explicated) invisibility might be connected to, more specifically might be an epiphenomenon of, His alphabetic origins. Certainly, there is little evidence within the West's theologico-philosophical tradition of the kind of sensitivity to the relation of alphabetic writing to depiction and more particularly to its interdiction that might have led to the posing of such a question.

Thus, consider a telling (but representative) twentieth-century example of the tradition's insensitivity: Edmund Husserl's celebrated essay "The Origin of Geometry." Geometry, as the study of spatial extension, is an inherently 'pictorial' domain of mathematics, in that—regardless of its various algebraic representations and axiomatizations—diagrams are inseparable, as noted earlier, from the ideas geometry studies and the means by which it furthers itself. Diagrams, however, are absent from Husserl's text. He neither employs them within his essay for purposes of illustration or explication, nor does he cite or explicate them as signs or make any comment about their special function and ubiquitous presence within geometry. He seems, in short, not to be aware that their presence might be a matter of any mathematical significance.[16] The phenomenon is by no means confined to Husserl. Jacques Derrida's extended two-hundred-page close reading of the essay (whose principal focus is no less than a philosophical examination of Husserl's use and theorizing of written signs) is likewise silent about the significance of diagrams vis-à-vis alphabetic texts as well as silent about Husserl's several silences.[17]

The Greek 'I': Psyche

As one knows, the role played by the written 'I' in Greek culture is most visible and copiously documented as a personal pronoun to be found in inscribing the ceremonial, political, artistic, and philosophical productions of human subjects that came to dominate the Greek development of alphabetic writing and reading. Unlike the syllabic writing it derives from, which requires much contextualizing work to decode any but the simplest inscription, the Greek alphabet is described as *phonetic*, is said to notate *all* sounds needed for the inscription of speech, making the project of transcribing and decoding of spoken language relatively transparent and hence facilitating the construction of a widespread literacy.

But the alphabet does not in fact notate all sounds of speech. It records 'words' (items which in fact owe their separate existence to its very mediation) but not how the words are spoken, not how speech is delivered, how vocalized, to its hearers. As we saw above (chapter 1), alphabetic writing omits the prosody of vocal utterance—the rise and fall, the tone, the rhythm, emphases, breaks, pauses, and silences of utterance—and thereby fails to record the different registers and multitude of meanings, attitudes, and affects discharged by these vocal gestures.

Recognition of the gap between speech and its inscription and re-actions to the falling short in affect and intended meanings that issued from it were an important vector in the development of Greek literacy and in particular the writing of 'I.' One of the earliest arenas in which this shortfall was confronted was public speaking, when a speech was delivered from a text by someone other than its author. Thus, for example, Gorgian rhetoric, concerned to redress the loss of prosody and transfer the force and affect of figures of speech to the written text, was born, according to Roland Barthes, when funeral panegyrics, originally composed in verse and spoken by bards, were "entrusted to statesmen . . . and if not written (in the modern sense of the word), at least memorized, i.e., in a certain fashion *set*." (1988, 18) And likewise in theatrical performance since, if, as Derrick de Kerckhove and others have argued, "Lyric poetry registered the first traumas of selfhood" (1981, 36), then it was perhaps inevitable that with the birth of a theatre of individual characters speaking lines written by another, the written but declaimed 'I' would turn on the relation be-tween lyric texts and their voiced delivery. In turn, questions of rhetoric and the persuasional possibilities and falsifications opened up by theatrical mimesis provided a written narrative form and critical—anti-sophistic—framework for Plato's writing down of Socrates's dialogues.

Before writing, in Homer's epics, figures attributed their actions to messages and commands received from the gods outside themselves. In the Odyssey, Penelope set up her loom to weave a robe "because some god breathed into my lungs that I should." (quoted in Olson 1994, 239) Socra-tes, he who did not (and perhaps could not) write but could hear writing's imprint inside the commanding voice of his godlike daemon, was the last of such oralists. By Plato's time the oral and aural gods as sources of advice and commands to a speaking 'I' were all but silent, their messages and outpourings, expressed in writing, had become 'thoughts' of a written 'I' originating from an invisible, disembodied 'mind.'[18]

Media create their most powerful effects when they efface evidence of their activity. Writing is no exception, so that no explicit—*written*—ac-count of the circumstances of such a ghost-agency's scriptive birth is to be expected, but semantic change over the period of writing's introduc-tion indicates that it took place. In oral Homer the words *psyche* or *nous* or *thumos*, all subsequently translated as 'mind' or 'soul' or 'spirit,' had no such mentalist, unembodied, or spiritualist meanings, but "depended to a large extent," according to Bruno Snell, on an "analogy with the physical organs," and *soma* denoted not a body opposed to a 'mind' but a corpse. (1960, 14) By the classical period *psyche* was for Plato an individual's mind-soul, and *nous* for Aristotle was the thinking principle. Only then do these terms "come to be united," as David Olson puts it, "in a general concept of mind as a mental organ in the head, and only then does mind come to be seen as contrastive to, and in control of, the body." (Olson 1994, 239)

There are two aspects of 'mind' here: its abstract status within Greek discourse as an incorporeal being in opposition to the concrete palpability of embodied flesh; and its putative agency—the mind as initiator, the generative source of ratiocination and 'author' of an individual's ideas and thoughts.

On the former, it is undoubtedly the case, as Eric Havelock, Marshall McCluhan, and many since have argued, that writing per se, the very nature and action of it as a medium, puts into play and maintains a cognitive leap in abstraction within which the concept of mind might make its appearance as an abstract entity. By extracting from speech the fundamental idea of a meaningless, monadic phoneme and segmenting the evanescent flow of spoken utterance into fixed, repeatable, examinable, theorizable, and context-free letter-strings notating phonemes, the apparatus of alphabetic writing and reading in itself, through the very conditions it imposes for its use as a medium, created the semiotic frame, psychological wherewithal, and conceptual matrix of Greek abstract thought. In addition, another source of abstract thought was the Greek invention of circulating money in the mid-sixth century in Lydia. Thus classicist Richard Seaford argues that the monetization of Greek society played a crucial role in the formation of pre-Socratic philosophy from Heraclitus onward by offering a material model for the division between objects of the senses and an abstract mental reality and hence the Heraclitean idea of an immaterial soul and ultimately mind. "Both monetary value and the mind are abstractions, embodied and yet in a sense invisible. Indeed each is a single controlling *invisible* entity uniting the multiplicity of which in a sense it consists." (Seaford 2004, 242) For money the unity is the interchangeability of all that can be assigned a monetary value; for mind it is "the formation of the unitary self out of internal fragmentation" put in place by Heraclitean monism.

But, despite their evident explanatory power, neither a generic, writing-induced abstract milieu nor a monetary-inspired abstract monism says anything about agency,[19] about the mind as an initiator of thought 'inside' a self, about the relation of self to the 'I'; and a fortiori about the specific enunciation that designates it. Once again the question of medium obtrudes: Is the 'I' spoken or written? For it is on this difference that the question of how mind, understood either as psyche-soul or as *nous*, the mental organ of thought, might have come to be in opposition to and in control of the *soma*. How does it come about that the mind can be "embodied and yet in a sense *invisible*"? The question of an invisible body (the obverse of a disembodied ghostly apparition) is ultimately a neurophysiological one. It asks: What does the alphabet require of the 'body' of its user? Better, how does the alphabet construct the kind of body/brain it needs in order to function? The answer, explicated earlier, lies in the fact that by eliminating prosody, severing the voice's gestures from the uttered 'words' whose letter-strings are their textual representations, the alphabet puts in place—enacts—a neurological hierarchy: the limbic midbrain, site of vocal affect intrinsic to the speaking 'I,' is subject to a cortical override. Its vocality is suppressed by the apparatus responsible for processing scriptive abstraction—words—shorn of all prosody.

At the level of the text this hierarchy is naturalized as ontological fact. In order for the discourse of alphabetic writing to imagine, as it does, that its texts convey thoughts about matters entirely external to itself, that texts are 'about' ideas and concepts unaffected or uninflected by their alphabetical transcription, to assume (even in principle) that texts could be independent of the effects of the alphabet as a medium, one must necessarily invoke the existence of a source, an origin of the thoughts it 'expresses,' situated outside the domain of the letter, an originating agency which writes but is not itself written. To write 'I' and to conceive what is involved in the process of doing so on the analogy of the 'I' of speech, as was (and is) the case, thus necessitated the putting into reality of an 'I'-er which replaces speech's location in the speaking body by an invisible, nonsomatic agency. Conflating the spoken and written 'I,' being unaware of any necessity to separate them, is precisely the means by which the scriptive coming into being of such thoughts is identified as the work of an invisible agency—*nous*—a disembodied ghost operating within the body of every thinking human being, a ghost infamously reimagined by Descartes at the onset of modern thought as *res cogitans*, as the author of his *written* 'utterance' *cogito*.

The Greeks created the Hellenic alphabet from the Phoenician syllabary by adding letters for vowels and some consonants. Their motives and their initial deployment of alphabetic writing are a matter of scholarly dispute.

Was it (as for the Phoenicians) technical and instrumental: to facilitate exchange and commercial activity? Or was it literary and artistic: in order to write down lyric and epic poetry?[20]

Supporting the former, one scholar observes that the majority of the earliest inscriptions, from the first half of the eighth century BCE, are about property, a prerequisite of exchange. The inscriptions are scratched directly on objects and "indicate ownership or the agent of manufacture" of the object on which they occur, either in the form of 'I or this belongs to X,' or 'X made me,' as in "I am the cup of Nestor," or in the form of a curse "I am the lekythos of Tateie—may whoever steals me be blind." (Ragousi 2001, 6) Other examples of early (circa 750 BCE) alphabetic inscriptions, funereal and commemorative texts inscribed on objects, such as "I am the kylix of Korakos" and "I am the commemoration of Glaukos," reveal a use of the written 'I' by inanimate objects concerned more with the question of memory and enunciation than ownership of property and trade.

What is striking about these very early Greek uses of the alphabet is the unexpected dimension of writing they indicate. They reveal that at its beginning the medium was invoking the extra-human, extending communicational agency and according reflexivity beyond human subjects to record—better, to institute—the 'speech' and claims of inanimate objects.

Thus writing is not only virtual speech in the sense of 'speech at a distance,' 'speech outside the milieu of its production,' but also virtual in the sense of 'speech removed from persons,' 'speech outside the human.' Obviously one can ask of these inscriptions, who spoke them? Certainly not the objects on whose surface they occur and whose presence they announce. Evidently "the words are real, *someone* wrote them, but beyond that the question doesn't even make sense,"[21] in that whoever scratched the letters is of minor significance and irrelevant to the status of the self-enunciation they contain. In a study devoted to the phenomenon of the inscribed 'I,' Jesper Svenbro observes about such writings upon objects: "These inscriptions are not transcriptions of something that could have been said . . . and subsequently transcribed upon the object. . . . Quite to the contrary: these statements are in some sense characteristic of writing, which allows written objects to designate themselves by the first-person pronoun, even though they are objects and not living, thinking beings." (quoted in Heller-Roazen 2005, 160) Indeed, as we have seen, *Jahweh*, an extra-human author of 'I,' a postulated 'I'-er who writes on a scroll of animal skin and declares Himself to exist within it, which thereby becomes a fetish for His invisible presence, operates precisely along self-designating lines.

In summary, as hypostatizations of the 'I'-effects that writing permits, Mind and God have functioned since the beginning of alphabetic inscription as disembodied, immortal, and invisible ghosts haunting thought about thinking and the nature of being in the West from the moment of their birth.

The Infinite Mathematical Agent

But let me move to a third ghost, one that mathematicians invoke, though they don't describe it so, when they write 1, 2, 3, . . . and think infinity.

A salient feature of ghosts is their mode of embodiment and disembodiment. Before all else, their aberrant or quasi-physicality underlies their strangeness, the spookiness of their presence, and inflects all inferences about them.

The medieval (Christian) king had, according to Ernst Kantorowicz, two bodies: a private and human one—his mortal, material body; and a divine, everlasting one. The monarch's presence, royal status, and exercise of power derived from the institutional, theological, and rhetorically assembled co-presence of these bodies. The mathematician has three bodies, or three material arenas of operation—a mortal Person, a virtual agent, and a semiotic Subject—likewise co-assembled. The mathematical *person* subjectively situated in language is the one who imagines, makes judgments, tells stories, and has intuitions, hunches, and motives; the mathematical *agent*, imagined by the Person, is a formal construct which executes ideal actions and lacks any capacity to attach meaning to the signs which control its narratives; and mediating between them as their interface, is the mathematical *subject*, who embodies the materiality of the symbolic apparatus that writes and is written by mathematical thought.

One can view mathematics as a thought-experiment process of Peircean 'reflective observation' according to which the person imagines the agent performing an activity and observes the result of the activity via the symbolic mediation of the subject. The agent is a proxy or surrogate of the person, so that for the observation to be a convincing thought experiment the agent must resemble the person. But the resemblance is necessarily partial: the agent is invoked to execute an action—such as unlimited counting—that goes beyond the person's temporal and/or material constraints. The agent is thus a person without a body, or rather a person with a virtual body that has a split character. On the one hand it lacks those features of bodies that prevent the person carrying out the action. On the other, any feature not so excluded remains unaffected and available to the agent.

Precisely this sort of specific exclusion and unlimited inclusion is, according to cognitive anthropologist Pascal Boyer, the characteristic of ghosts that figure in religious settings. The mathematical agent and the inferences that can be made about it parallel those of supernatural entities and their inferences. In each case the governing logic arises from the same double move: "Religious concepts *violate* certain expectations from ontological categories [and] *preserve* other expectations—[namely,] all the relevant default inferences except the ones that are explicitly barred by the counterintuitive element." (Boyer 2001: 62, 73)

For Boyer, the ontological categories here have a biological genealogy. They comprise natural, that is, pre-linguistic, evolutionarily determined, unconsciously mobilized, and instantly available templates. There are a small number of these—namely *Person, Tool, Animal, Thing, Plant*—each the source of multiple, readily produced, and habitual expectations. The double move he outlines results in concepts that are at once un- or non- or super-natural—precisely the result of the violation—and yet highly productive and stable on account of their remaining un-interdicted wealth of default inferences. A traditional ghost, for example, might pass through walls, thus enjoying what the linguist Leonard Talmy identifies as "fictive motion" (Talmy 2000), but be able to see and hear with unimpaired human powers.[22]

For classical mathematics the ghost activity of interest is infinite counting, and the relevant ontological category that is violated is that of person. Expunged are all expectations and inferences that spring from the person's physicality—fatigue, mortality, boredom, bounded inscriptional resources, imperfect repetition—which militate against any attempt to iterate indefinitely. However, as I've argued elsewhere (Rotman 1987/1993) the supernatural or ghost character of the classical agent is not evident, since it is masked by the prior naturalization of the endless sequence 1, 2, 3, . . . of so-called *natural* numbers as an unexamined given, "a gift from God," definitional of mathematical thought itself.

This prior naturalization is, in effect, built into the logic Aristotle wielded to discuss the nature of counting. It presupposes the capacity—taken as unquestioned, implicit, and obvious—to be able always to count one more time. As such it has since become the rhetorical and conceptual basis for the potential infinite in mathematics. Potential, that is, against the paradoxical effects of the *actual* infinite threatened, in Aristotle's understanding of them, by Zeno's paradoxes.

The agent of endless iteration, then, is *nous*, the immortal disembodied thinking organ deployed as an active mathematical principle. An extension of it, the repudiated actual infinite, re-entered mathematics in the sixteenth century within the theory of infinite series, and again more radically in the nineteenth century. At the moment Friedrich Nietzsche was announcing the death of God, Georg Cantor was transposing a version of Him into a mathematical principle. The violation this time concerned the ontological category of Thing. The part–whole assumption about any collection of 'things'—the whole is always greater than the part—seemingly inherent to the very idea of a collection, an assumption whose failure had so troubled Galileo's reflection on the endless sequence of numbers, was converted by Cantor into a definition. As a result what had been a horizon—the outer limit to *nous*-imagined human counting—became an object of mathematical discourse. And the frankly theological concept that Cantor called 'the Absolute,' with its paradoxical and inaccessible set of all sets, took the place of this displaced limit. The result was an agent with the capacity to enter into narratives about infinity itself.

Cantor's infinities, together with their set-theory matrix and supporting Platonist theology, exerted a hegemonic influence over the presentation and much of the content of twentieth-century mathematics. Recent developments, however, indicate that the days of its unquestioned dominance are numbered. Digital mathematics—intrinsically antagonistic to infinitary thought and already the basis of a new experimental mathematics of computed objects—would appear to be its nemesis.[23] Where traditional mathematics works through proofs about imagined objects—the narrative actions of a weightless ghost—digital mathematics simulates these as real-time, material procedures of a machine. This, as elaborated in chapter 3, has various consequences for the content, form, and overall development of mathematics, particularly in relation to the concept of infinity. In short, the traditional syntax-driven mathematical discourse of symbols, notation systems, and formulas understood axiomatically and organized into linear chains of logic reducible to pictureless first-order languages, is confronted by a discourse that is performative and experimental and driven by digital, screen-visualizable images for which proof and logical validation, while by no means absent or unimportant, no longer dominate or have priority. This is not to say that the classical, infinitary agent will disappear, or be legislated as 'invalid,' or be repudiated as an illegitimate or illusory metaphysical concept, but rather its ideality, the supposed 'naturalness' of its ghost ontology, cannot but be revealed and ineluctably reconceived when confronted by the materializing, de-infinitizing action of digital computation.

Networked 'I': Future Ghosts?

What sort of ghost might the contemporary technologies of the virtual be nudging into existence? If the agencies facilitated by speech and writing are a guide, the site for its emergence will be a self-enunciation proper to networked media. But the question is surely premature, since such a new, necessarily post-alphabetic, reflexivity has yet to arrive. One can nevertheless observe that the current state of the alphabetic 'I' is one of increasing dissolution, as digital media and their successors continue to undermine and reconfigure the alphabetic matrix that holds it together. In this ongoing upheaval the old mono-self, the one-thing-at-a-time, linearly progressing lettered psyche with a sequentially orderable memory and a timeline history, is disappearing. Or rather the hegemony, undisputed authority, and automatic intellectual and spiritual preeminence of such a writing-engendered monad is diminishing, giving way to a para-self, a parallelist extension of the 'I' of alphabetic literacy that is crystallizing around us. And with it the conception of a single Truth and a singular notion of 'truth,' a single theory of the world, a single all-encompassing deity, become increasingly difficult to sustain.

This opening up and coming apart of a psyche that had so long narrated itself as a self-sufficient, autonomous 'I' held within the monadic confines of alphabetic writing, has a techno-somatic correlate. Neurological implants, vat-grown organ transplants, and robotic prostheses are increasing the body's externally derived content and functioning. Genetic, cellular, pharmaceutical, psychotropic, and biomedical augmentation are altering the body's underlying software. Machinic intervention through the technologies of gene analysis, brain mapping, body scans, and internal scopic procedures, are breaking down the boundary between inner and outer knowledge and control of the body. The result is a body which, though conditioned by and inseparable from its evolutionary lineage, is revealed as increasingly exogeneous — made and conceived from its bio-techno-cultural environs; increasingly transparent — less privately enclosed, more publicly inspectable and surveyable through a multitude of techniques; increasingly porous — engaged in a constant flow of information and affect

across its boundaries; increasingly heterotopic—an assemblage of different processes with their own histories, dynamics, and itineraries understood collectively, conceived as "a type of world full of an infinity of creatures." (Deleuze 1993, 109) It is also, as we've seen, more recordable and
writable through new forms of body graphism such as technologies of
motion capture which allow the gestural mobility of the body to be mediated—detached, rerouted, and realized anew—in a widening array of instrumental, aesthetic, and artistic contexts from telesurgery to virtual theatre. Not coincidentally, at the same time as the body's tactility and visible
movements are becoming digitally mediated, our modes of interface with
communicational networks are being articulated, experienced, and performed in gesturo-haptic terms of touch, movement, and immersion. The
"dominant sense in this world of pervasive proximity is no longer vision
. . . [but] touch." (Federman 2005, 8) And, as if to emphasize the alphabet's
attachment to visible letters and its aversion to gesture, Derrick de Kerckhove observes that the Internet "is not really amenable to sight as much as
to touch. Navigating the web is a tactile affair." (2006,)

It would be surprising, given it is a metonym of and production by
the body, for there not to emerge a corresponding self, the creation of
an 'I' in tune with and experientially appropriate to such embodiment.
A psyche that is at once porous, heterotopic, distributed and pluralized,
permeated by emergent collectivities, crisscrossed by networks of voices,
messages, images, and virtual effects, and confronted by avatars and simulacra of itself. In short, a *para-human* agency which experiences itself as an
'I' becoming 'beside itself.' Such an 'I' could in no way be an alphabetic
construct. Alphabetic letters, chained to the monad and the line, never
escape their one-dimensional seriality, and so are intrinsically antagonistic
to parallel processes, not least to the capture or representation of a simultaneity of heterogeneous events. Earlier we saw how, on the contrary,
digital imaging along with parallel computing were media able to present
and metaphorize a multiplicity of co-present actions. Parallel computing
is in a sense a subjecting to digital control of old and familiar deep-seated
biological and social processes, from multi-cellularity to the division of
labor, and its effects on subjectivity are correspondingly profound. By distributing an individual linear consciousness, a monadic thinking self, over
a collectivity, its action both pluralizes the alphabetic 'I' behind this consciousness and correspondingly reconfigures the social multiplicity, the
'they/we' against which it is defined.

A more specific computational metaphor for such an 'I' comes from the recent development of quantum—post-digital—computation. A core quantum phenomenon is superposition: the counterintuitive and seemingly paradoxical feature of any quantum system to be in a multiplicity of different virtual states at the same time, a multiplicity which collapses to produce a single, actual state as soon as the system is observed and/or measured. Though in its infancy as a practical source of parallel computation (and hence as a source of psychic parallelism), its phenomenological relevance to the emerging self that I've been sketching is suggestive: a quantum self would be one that exists as a co-occurrence of virtual states, an 'I' which becomes actual or 'realized' and fixed as an experienced and 'objective' whole precisely when it is observed, subjected to psychic measurement or social control, or otherwise called upon to act, respond, be affected, and project agency. Such an 'I' would be a mass of tendencies, an assemblage in a perpetual state of becoming, rather than a monolithic being, an identity "in perpetual formation and reformation at the moment of use."[24] (de Kerckhove 2006)

But, metaphors and heraldings of it aside, a stable self-enunciation—robust, iterable, recognizable—appropriate to the contemporary scene, an 'I' proper to the technologies of the virtual, has yet to emerge. For that to happen the ongoing recalibration of space-time and reconfiguration of agency by contemporary media must cease expanding their reach, cease altering the conditions of their own use, requirements which would seem to imply a plateau, a terminus or culminating point to their technocultural transformation. One possibility for such a limit is an internal exhaustion, the experience of diminishing returns coupled to global saturation. Something along these lines is suggested by Bernhard Steigler: "The development of technologies of the virtual is intensifying and accelerating a process of de-territorealisation, displacement, de-realization and disincarnation" effects which constitute, he says, "the ultimate stage toward the unification of the planet."(2001, 137). Only then, perhaps, when the network-induced perturbations of the space-time of their users have stabilized, can a unified, planetwide subject position, an electronically mediated enunciator able to refer to itself via the unified planetary network, become available. And then, in addition to this, only when its mediation has become invisible can a network-induced ghost or ghosts appear.[25]

There is however nothing immediate, automatic, or predetermined in such a process. Recall that for the medium of alphabetic writing (in both its Jewish and Greek forms), it was only after a protracted period of its use to write and read 'I,' along radically different itineraries, that the reconfigurations of agency set in motion by written speech had become implicit, only when writing had ceased to modify the milieu in which it operated and could conceal its mediation, did the ghosts we know as God and Mind achieve a stable presence. A medium-enabled ghost-construct will not become 'real,' will not in other words induce belief that it 'exists' and does so independently of human activity, until the medium's role in its birth has been effaced.

To finish, let me descend from speculation about future ghost effects to material history and return to God, as it were, to the purely physical substrate necessary for the writing of and believing in the Judeo-Christian God, to what Regis Debray, who embeds the issue in a more wide-ranging mediological account, designates as "theo-graphy."²⁶ One can identify three remediations decisive in the written development of the Judeo-Christian Being's presence and his worship.²⁷

First, inauguration of God's written presence: the switch from the cuneiform incising of syllables on small, hand-sized clay tablets to writing alphabetic letters on skin, which allowed more extensive texts to be produced, ultimately the many-skinned *Torah* and the grand, linear narrative its scrolled form demanded. Second, inauguration of Christianity: the switch from an enforced linear reading to the random access facilitated by bound codices, allowing one hand to hold the codex and the other to point elsewhere (something not possible with a scroll) and enabling the isolation, quotation, and rhetorical juxtaposition of passages essential to Christianity's construction as a proselytizing religion of the reported, transportable word. Third, the inauguration of Protestantism: the shift from the handwritten, error-prone copied codices to the mass-produced printed inerrant book—the Vulgate Bible—and the universal, directly accessible privatization of the word it made possible.

And now the emergence of a fourth remediation in the move from the paper materiality of printed books to electronic writing on the networked screens of digital media. The obvious question: What might be the accompanying discontinuity, the fourth media-induced break in the tradition of Judeo-Christian monotheism? A remediation pertinent not just to Judeo-Christianity, but to the status of *any* religion authorizing its transcendental status by means of a holy alphabetic text, which would of course include Islamic monotheism.

If such is the case, if changes in the material base, in the writing and reading wherewithal of worship and belief practices, are cognate with changes in the theological and phenomenological superstructure, then one can read the rise and frenzied appeal of Bible-obsessed evangelism and the fundamentalist surge in Jewish and Koranic literalism as reactions—fearful and defensive—to a perceived threat, nothing less than the end of the writing-based era which gave birth to and has thereafter circumscribed them: the threat of a God in danger, a God displaced, a God about to be obsolesced by the heathen and secular presences that electronic technologies seem to be conjuring into existence. To say this is not to suggest fundamentalists and Biblical literalists understand the threat as linked to the fate of alphabetic writing. Neither they nor monotheists at large concern themselves with the status of the alphabet or could entertain, even in theory, a mediological origin of their God—to do so would not only undermine their faith and nullify the Biblical and Koranic justification for their long-standing support of patriarchal repression, but would confront them with the inherent violence, inescapable and endemic, issuing from the claim that *their* God—Jewish, Catholic, Islamic, Protestant, Mormon—is the one, true monobeing.

For nonbelievers in the extra-human origin of the monobeing who are prepared to accept that alphabetic writing equipped Western culture with its absent God and Mind-soul, it becomes possible to go beyond the horizon of these two disembodied hosts. It becomes possible to imagine the end of the entire tradition of such writing-induced metaphysics, to perceive the archaism of these specters, their falling into disuse; possible to recognize we are approaching a particular moment in the history of Western writing—the beginning of the end of the 'era of alphabetic graphism.' It becomes possible, in other words, to think that the very concept, discourse, reality, affect, and persuasive hold of such entities will not survive the ongoing dethronement of alphabetic culture; possible to recognize that a dead end has been reached; possible to feel that the times are bidding us to shuck off these ancient, no longer helpful ghosts to make way for new ones, to make room for other less imprisoning, more open-ended, diverse, ecophilic, and planet-mindful ontological and ethical horizons of the human.

Possible . . . but, meanwhile, here in the alphabetic, all-too-archaic present, science in the form of physics dreams of a 'God particle' and pursues a monadic Grand Theory of Everything; a God-saturated America in thrall to the Bible remains convinced of its exceptional and special relation to the monobeing; and Muslims fight holy wars against infidels who dare insult God's one true prophet.

Notes

1 For reports on the progress in the area of brain-machine interfaces, see the special section of *Nature*, 442 (7099), 13 July 2006.

2 In many ways the argument grows out of an extended, appreciative but ultimately critical dialogue with Derrida's theses on writing and grammatology in the formation of the (Western) subject. While deconstructing notions of presence and logocentrism at the foundations of Western metaphysics, Derrida, like most other major Western philosophical thinkers, has still left us with a subject as disembodied, a floating signifier with no traction for agency.

3 Also see Nigel Thrift, "Electric Animals: New Models of Everyday Life?" *Cultural Studies* 18 (2/3), March/May 2004: 461–82.

4 For a discussion of "Baldwinian evolution" in relation to Darwinism, see Robert Richards, *Darwin and the Emergence of Evolutionary Theories of Mind and Behavior* (Chicago: University of Chicago Press, 1987).

5 We are not just a species that uses symbols. The symbolic universe has ensnared us in an inescapable web. Like a "mind virus," the symbolic adaptation has infected us, and not by virtue of the irresistible urge it has instilled in us to turn everything we encounter and everyone we meet into symbols, we have become the means by which it unceremoniously propagates itself throughout the world.

6 Elsewhere, Deacon summarizes his point eloquently as:

> I do not suggest that a disembodied thought acted to change the physical structure of our brains, as might a god in a mythical story, but I do suggest that the first use of symbolic reference by some distant ancestors changed how natural selection processes have affected hominid brain evolution ever since. So in a very real sense I mean that the physical changes that make us human are the incarnations, so to speak, of the process of using words. (1997, 322)

7 This position was originally championed by Lev Vygotsky in his work on mimesis in children in 1934 (see Vygotsky 1986). It has recently been reinvigorated by the work of cognitive scientists such as Donald and McNeill.

Preface

1 In a "HyperMail" posting of 26 May 1999 in the *Math Archives* (http://sun
   site.utk.edu/math_archives/.http/hypermail/historia/may99/0210.html), Walter Felscher of the University of Tuebingen believes that the earliest reference
   to Kronecker's dictum was in Kronecker's necrologue by Heinrich Weber in
   Jahresberichte D.M. 2 (1893) 5–31.

Introduction

1 A sensory modality whose abstract status falsely isolates it from the body requiring that it be reconnected, according to Gilles Deleuze's reading of Berg-
   son, with the body's movement and as a result supplemented by haptic and
   tactile functions (1988, 492–93). Compare Shaviro's (1993) implementation
   of these ideas to emphasize the visceral nature of the cinematic image, and
   Hansen's Bergson-inspired claim (2004) that it is only in the body's sense of
   movement, touch, and proprioception that 'seeing' can achieve its true corporeal dimension and so be understood.

One. The Alphabetic Body

1 Albeit different versions of 'the' alphabet. See the excellent Omniglot site www
   .omniglot.com for examples and discussions of different alphabets, abjads, syl-
   labaries, etc.

2 And will have different practical implications, particularly for the kinds of cognition and the conceptual affordances they facilitate. Thus for Barry Powell,
   the crucial innovation of Greek writing was less the vowel/consonant question
   than the cognitive and theoretical consequences of the very idea of a phoneme:
   "The revolutionary feature of the Greek alphabet was not the introduction
   of vowels, which are common in earlier writings, but the isolation of graphic
   signs that represent phonemes in natural language." (1994, 6)

3 My interest in gesture here is chiefly its relations to speech and writing. For a
   discussion of the importance of gesture and embodiment with respect to oral
   poetics, see Zumthor 1990, especially chapter 10.

4 The full title of which is *Chirologia: or the Naturall Language of the Hand. Com-
   posed of the Speaking Motions and Discoursing Gestures Thereof. Whereunto Is
   Added Chironomia: or The Art of Manuall Rhetoricke. Consisting of the Naturall
   Expressions Digested by Art in the Hand as the Chiefest Instrument of Eloquence.*

5 The claim that the origin of syntax in spoken languages is to be found in the
   intrinsic properties of a gestural act is explored in Armstrong et al. 1995.

6 The point deserves considerable elaboration. Thus the obligatory gestures and
   body practices that constitute Islamic worship and indeed define what it means
   to be a Muslim are far more pronounced than anything resembling them in the

other Abrahamic religions and seem, in terms of the inculcation and maintenance of fervent, unshakeable belief, to be correspondingly more effective.

7 The same claim, but not tied to religious expressivity, threads through the word–image opposition insofar as this turns on the evocative and affective power and belief-maintaining capacity of visual emblems—badges, logos, flags, posters, insignia, icons—over texts.

8 I have taken these examples from various sources, principally McNeill 1992 and Martinec 2004. The latter contains a useful survey and elaboration of the experiential dimension of gesticulatory gestures which, Martinec claims, contrary to Kendon 1972 and McNeill, are compositional and not holistic.

9 Of course, the body produces a wide range of acoustic events, such as grunts, squeals, chortles, sighs, screams, farts, cries, moans, sneezes, exclamations, burps, whistles, boos, laughs, snorts, sniggers, and others, many of which are, or are able to be, motivated with social meaning and semiotic content; content that frequently duplicates or overlaps or on occasion subverts the regulatory, expressive, and communicational functions performed by speech and by emblem gestures.

10 From a different direction, see Speer et al. 1993 for an account of the crucial role prosody plays in the memory and recognition of phrases, sentences, and musical passages.

11 For an account of the initial impact and significance of the typewriter on literary practice, notably in the style of Friedrich Nietzsche's later writings and Franz Kafka's letters, see Kittler 1999.

12 Tedlock 1983, chap. 1, provides an extensive examination of the gap between speech and its inscription, from an ethno-poetical perspective framed precisely in terms of the elimination of vocal gesture and the body movements that affect voice.

13 The model in question is the so-called triune brain consisting of three, evolutionarily successive layers: the ancient reptilian brain stem, the mammalian midbrain, and the more recent cortical (especially neocortical) uppermost layer. See Wilson 2004, 84–87 for an elegant articulation of the utility of the model despite its acknowledged oversimplification.

14 Mathematical ideograms constitute the most extreme development of such cortical autonomy. The fact that mathematical languages are devoid of any reference to an 'I,' or to an *other*, that is, to an embodied writer, listener, addressee, or signee, and are at the same time the ur-site of a supremely abstract and disembodied ontology, should not therefore be surprising. See Rotman (2000, 15 and passim) for more.

Two. Gesture and Non-Alphabetic Writing

1 The model of mathematical learning and thinking indicated here is part of a larger and more diverse ecological framing of abstract thought. See Wise 2004

for a series of contributions to what might be called the 'ecological turn' in cultural studies of science and technology.

2 See Rotman 2000, chap. 2, for a fuller account and critique of this dismissal of diagrams as epiphenomenal or mere psychological props, supposedly eliminable in principle from formally 'correct' and 'rigorous' mathematics.

3 In his introduction to Châtelet's essay, Kenneth Knoespel (2000) lists the major features Châtelet ascribes to diagrammatic operations as: Diagrams constitute technologies that mediate between other technologies of writing. Diagrams create space for mathematical intuition. Diagrams are not static but project virtuality onto the space they seek to represent. Diagrams represent a visual strategy for entailment. Diagrams are mediating vehicles, which means they cannot only be recovered but also rediscovered. See also Knoespel 2004 for an exploration of a possible 'diagrammatology,' part of the "far more articulate anthropology of scientific discovery" opened up by Châtelet's essay.

4 Interestingly, if one examines mathematics from a semiotic perspective, then the experience of doing and writing mathematics as a symbolic activity can be encompassed within a thought-experiment model first sketched by C. S. Pierce. According to this model, assertions made by mathematicians are revealed to be predictions, foretellings of the mathematician's future encounter with signs. A brief description of the model is given in chapter 3. For a fuller account, see Rotman 1993 and 2000, chap. 1.

5 See Wilson 2004, especially chap. 5, for an extended exploration of what might constitute 'gut feelings' in terms of the connection between neurology and affect not subsumed by the central nervous system. For a comprehensive examination of the evolutionary background of gut feelings and the non-conscious and/or pre-rational decision making allied to them within human thought and action, see Gigerenzer 2007. For a popular account of such decision making articulated in terms of the particular method of 'thin slicing,' see Gladwell 2005.

6 Some examples of mousegrams can be found at http://www.wideopenwest .com/~brian_rotman/mouse-display.html.

7 Or 'Performance animation' to use the term suggested in Furniss 2006 on the grounds that 'motion capture' places undue emphasis on the technology of the process as opposed to its functioning as a medium. Other suggested names she cites for the medium (each with its own emphasis) are 'performance capture,' 'virtual theater,' 'digital puppetry,' 'real-time animation,' and, a favorite with designers apparently, 'devil's rotoscope.' In what follows I shall stick to 'motion capture.'

8 Two other forms of motion or gesture 'capture'—transductions from gestural to other media—deserve to be mentioned here. The musical instrument known as the Theremin, the first purely electronic instrument and the precursor of the Moog synthesizer, converts hand gestures in the neighborhood of its antennas into musical notes; and the design method of 'air sculpture' which uses three dimensional movements of the hands (monitored by cameras) to create a vir-

tual object in model space which can then be formed into a real object whose contours reproduce those carved out by the original gestures. Though neither of these methods of capture constitutes writing in any narrow sense, they both effect transductions from the medium of gesture to those of sound and shape production which bear suggestive mediological comparisons to the transduction into text that writing performs on the medium of speech.

Three. Technologized Mathematics

1 For an extended treatment of this issue see Rotman 2000, chap. 2.
2 The difference here between testing and finding a solution or, what can be the same thing, recognizing and producing one, has correlates outside the field of mathematics and computing. In everyday life one often can recognize a phenomenon when confronted with it—such as beauty or worth—without being able to explicitly characterize of define it. Similarly, St. Augustine knew what time was, he could recognize it when it was present to him but had no way of explaining it, no way of arriving at a characterization of it that would be communicable to others. *What then is time? If no one asks of me, I know; if I wish to explain to him who asks, I know not.*

Four. Parallel Selves

1 For examples: Foucault's elaboration of self-scrutiny as an internalization of the confessional and more public technologies of panoptic surveillance; McLuhan's identification of the psychic constrictions of print technology (1962); Jean-Louis Comolli's work on "machines of the visible" linking filmic materiality to audience perception (1980); Kittler's tracking of the triple usurpation of the book as the chief imagination machine of Western culture by the media of film, typewriter, and gramophone (1989); Havelock (1963), Ong (1982), Goody (1977), Harris (1995), Olson (1994), and Eisenstein (1983) on writing's and printing's restructuring of speech, cognition, and consciousness; Latour's tracing of nonhuman agency (1993); Lenoir and Sha's articulation of bio-medical subjectification (2002); Lyotard's self as "located at 'nodal points' of specific communication circuits" (1984, 15); and many others.
2 This is particularly evident when the technology is seen primarily in semiotic terms as creating certain cognitive subjects—those who 'use' or 'practice' (i.e., master and are mastered by) pre-given codes. See Rotman 1987/1983 for an elaboration of the phenomenon in the codes of arithmetic, painting, and finance in terms of the repeated activities and routines performed respectively by the one-who-counts, the one-who-depicts, and the one-who-transacts.
3 See Stenning (2002), especially chaps. 2 and 4, for a detailed juxtaposition and analytic comparison of depictive ('diagrammatic') and textual ('sentential') modes of thought from a cognitive science perspective.

4 Two remarks. One, the anecdote of the clocks: a man heard the clock strike two times one day as he was falling asleep, and he counted: "One, one." Then, realizing how ridiculous that was, observed, "The clock has gone crazy: it struck one o'clock twice!" (quoted in Ifrah 1985, 24) Two, synthetic precedence: Kuratowski's definition, crucial in the formal theory of sets, of the ordered pair (a,b) constructed out of the unordered set {a,b} of two objects, and its propriety discussed by mathematical logicians.

5 Or so I argue (Rotman 1993): the ordinal/cardinal split is articulated as the difference between what I dub 'iterates,' i.e., ordinals counted into being, and all other numerical multiplicities, 'transiterates'—essentially cardinals that cannot be so counted.

6 This is not to say that there were not linear problematics that were in need of solution, thereby consolidating the prior linearity of the conceptual apparatus. Thus navigation and warfare gave the mathematics of celestial mechanics and ballistics central place in the mathematical agenda of the seventeenth century. And these mathematical concerns were amenable to being treated by sequences of calculations that begat calculus; a cognitive technology that, notwithstanding its focus on change in space and its apparatus of diagrams that go with it, is a linearizing mode of thought par excellence.

7 For more on linearity, especially in relation to the real number line, see Rotman 2000, 75–78.

8 There are numerous works devoted to what are variously called the new or digital media that provide useful analyses. Three of interest to the thematic here are Manovich 2001, Stephens 1998, and Hayles 2005. The first is a survey of specific effects and affects achieved and occluded by digital apparatuses and their protocols across the entire media field; the second charts conceptual, informational, and affective facilitations, antagonistic to print culture, which arise from editing techniques deployed by the 'new' post-filmic video; the third is a comprehensive analysis of media effects seen in terms of their production, storage, and transmission capabilities elaborated and exemplified through close readings of science fiction literature.

9 See Batchen 2001, chapter 6, for a collection of pronouncements of photography's death in the face of digitality, together with an analysis of the anxieties that drive them.

10 A similar anti-metaphysical lesson can be drawn from the use of digital imaging tools in mathematics which foreground the existential status of mathematical entities by confronting mathematicians with the juxtaposition of two opposed understandings of the objects they study. The classical, orthodox viewpoint: mathematical objects are transcendental, invisible, and imagined versus the digital, understanding of them as materializable, variously idealizable, and imagable. The difference is fundamental: the first describes, for example, a Euclidean point, contentless, infinitistic, and zero dimensional; the second, a material pixel with real, specifiable dimensions and variable information content.

11 According to Merlin Donald, the development of external memory and thinking devices central to the evolution of cognition achieves a radically new level
in the case of humans: the sheer plasticity of our brains ensuring our immersion in technology will constantly reconfigure our neural connections, which
means that, just as alphabetic writing reconfigures the brains of its users, so
these visualization apparatuses, more than mere enabling devices, are rewiring
the individual viewing brains and minds that—collectively—imagined them.

12 For example, work by the British artist and photographer Idris Kahn whose
single-image composites of pre-digital photographs (by Karl Blossfeldt, the
Bechers, Eduard Muybridge), paintings (by Caravaggio), musical scores (by
Bach, Mozart, Beethhoven), and entire books (the *Koran*, Freud's *The Uncanny*, Sontag's *On Photography*, Barthes's *Camera Lucida*) offer a literal, direct,
and at times eerily arresting realization of a parallelist optic.

13 The personalities can be differently related to each other: in some cases a dominant and a variety of subsidiaries; in others, alternating dominants; most claim
to be solo, but some are aware of their co-inhabitants; some are fully worked-
out personae, others personality fragments and generic functions, such as 'the
Angry One,' 'the Innocent Child,' etc.

14 For example, the experience of Robert Oxnam as narrated in his memoir
(2005).

15 A response to the disorder in accord with Hacking's iatrogenic approach claims
that the sufferers have a condition that is narrativized into existence by the
present-day therapists who treat them (Acocella 1998).

16 Compare the claim by Pierre Levy (1997, 1) who urges that "toute une société
cosmopolite pense en nous" (the whole of a cosmopolitan society thinks within
us) and his Darwinian formulation of a psyche at once individual and collective: a "collective subject . . . multiple, heterogeneous, distributed, coopera-
tive/competitive and constantly engaged in self-organization or autopoeisis"
(1998, 128).

17 See N. Katherine Hayles (1999) for a critique of the projection of the post-
human based on the idea of purely abstract, disincarnate information as providing the means for a longed-for escape from the situated and embodied necessities of human thought.

Five. Ghost Effects

1 And it ignores the growing awareness among animal ethologists that certain
mammals exhibit forms of a-linguistic thought such as metacognition—"the
capability to reflect upon something that is mental not in the environment"—
as well as mirror recognition of (in a yet to be determined sense) them*selves*.
(Phillips 2006, 31)

2 At this point a terminological disclaimer is necessary in relation to a philosophical concept of the virtual elaborated by Gilles Deleuze. According to

this, and I unavoidably simplify and perhaps traduce, the virtuality of a being (which is always an assemblage in a process of becoming) concerns what is or was actualizable or realizable in it, what its lines of flight could or might have given rise to, the futures it could inflect or take part in. On this view, unlike potentiality, which refers to determinations that are fixed but lack the conditions to realize them, the virtual is not determinate; on the contrary, it is inseparable from tensions, problems, sites of instability, and responses to open questions always present in an assemblage. Accordingly, it would seem that what is actual or 'real,' or has been actualized or realized, is not, or can no longer be considered as, virtual. But this conflicts with contemporary usage — the virtual X that figures here — which is the basis for the concept I extract from the term. Thus, though virtuality in this sense does indeed separate something from the actual — from the X that is virtualized — it nevertheless names a new, indisputably actual and 'real' object, experience, or form of activity.

3 Compare the formulation by Merlin Donald: "The principle of similarity that links mimetic actions and their referents is perceptual . . . best described as implementable action metaphor." (1998, 61)

4 More fully: "The mastery of the first-person pronoun parallels the 'construction' of oneself as a self-conscious individual. . . . [One's] knowledge of oneself is the knowledge of oneself as a *social agent*." (177)

5 "If our personality survives, then it is strictly logical and scientific to assume that it retains memory, intellect and other faculties and knowledge that we acquire on this earth. Therefore if personality exists after what we call death, it is reasonable to conclude that those who leave this earth would like to communicate with those they have left here . . . if we can evolve an instrument so delicate as to be affected, or moved or manipulated . . . by our personality as it survives in the next life, such an instrument when made available ought to record something." Edison, quoted in *Scientific American*, October 1920.

6 "Electronic Voice Phenomena (EVP) concerns unexpected voices found in recording media. It is a form of after death communication. ITC is a newer term that includes all of the ways these unexpected voices and images are collected through technology, including EVP." American Association for Electronic Voice Phenomena Web page at http://www.aaevp.com.

7 These effects were imagined in contemporary works by Rudyard Kipling, Thomas Mann, Judge Schreber, Upton Sinclair, and many others. Kipling's 1902 story "Wireless" (1994) is particularly explicit. In it he explains how the "magic . . . of Hertzian waves" being then deployed by Marconi depends on induction: the magnetism of an electric current inducing a current in a wire unconnected to it. The tale concerns a coming together across time and space between a consumptive apothecary's assistant and the spirit of Keats. Waiting for an incoming wireless transmission, the narrator observes the assistant (in a drugged state) writing to his beloved Fanny Brand and then intoning and writing fragments of prose and verse which converge on Keats' poem "St. Agnes'

Eve." Meanwhile, instead of the expected transmission what is picked up are Morse code transmissions between ships trying unsuccessfully to contact each other, resembling a "spritualistic séance . . . odds and ends of messages coming out of nowhere." Kipling's mirroring—the common presence of consumption, Keats's experience as an apothecary, Fanny Brawne/Fanny Brand—sets up a psychic resonance, making a wireless-like transmission of the ghost of Keats possible, since "in conjunction with the mainstream of subconscious thought common to all mankind," the assistant has temporarily experienced "*an induced Keats.*" See Lecercle (2002, 202–17) for an extended reading of "Wireless." As indicated, Electronic Voice Phenomena, constitutes a contemporary recrudescence of radionic contact with the dead, enabled through the software analysis of background noise.

8 Among early writing, Henri-Jean Martin observes, "We can decipher funerary inscriptions everywhere [throughout Mesopotamia] in which the dead . . . ask that their names be pronounced or that an offering prayer containing their names be read aloud, almost as if that could make them live again." (1994, 102–3) As Friedrich Kittler remarks, "The realm of the dead is as extensive as the storage and transmission capabilities of a given culture." (1999, 13) And there is the "Legend of frozen words," already old in Roman times, as told by Plutarch: "In a certain city, words congealed with the cold the moment they were spoken, and later, as they thawed out, people heard in the summer what they had said to one another in the winter." Douglas Kahn cites this tale within a discussion of "enduring speech cloaked in a phase of inaudibility" and treats it as a myth finally fulfilled by the printed book. But it is surely also an echo, a memory in fable form, of the first impact of writing's wondrous and magical power to make speech live again. (Kahn 1999, 204)

9 In *Moses and Monotheism* Sigmund Freud constructs his narrative of the origin of monotheism and the connection between Akhenaten and a supposedly Egyptian Moses on historical data no longer accepted as correct. See DiCenso 1999, 79–80 for some expressions of this criticism. Contrariwise, Richard Gabriel (2002) argues at length for a direct link between Akhenaten, Moses, and the Jewish monogod.

10 The contemporary formulation of this understanding of the Biblical text, known as the Documentist Hypothesis, holds that the *Torah* is composed of four different textual currents by four different authors or schools, with the most ancient dating from around 900 BCE. See Finkelstein and Silberman 2001; Friedman 1987.

11 For Adriana Cavarero vocality and the uniqueness of each human voice is the founding principle for an entire anti-metaphysical project of ethical and political ontology, starting from "The simple truth of the vocal, announced by voice without even the mediation of articulate speech, communicates the elementary givens of existence: uniqueness, relationality, sexual difference, and age. . . ." (2005, 8)

12　This existential oneness is part of a cascade of co-occurring monoids: a single Aaronide priesthood victorious over the rival Mosaic faction; a single holy place, Jerusalem, instead of a plurality of sacrificial sites; a single Talmud told by a single author rather than a scattered archive of heterogeneous texts; a scroll whose beginning-to-end unrolling delivers a linear 'history' of a single people. The result of this assemblage of singularities was a monomania of self-hood and auto-individuation in which the *Jahweh* and the Jewish people, transcendentally exceptionalized through its everlasting covenants with *Jahweh*, simultaneously create each other.

13　For Louis Althusser this is a prime and extreme example of the mechanism of hailing or interpellation whereby ideological apparatuses produce a 'free' subject who is subjected, "stripped of all freedom except that of freely accepting his submission." "Moses, interpellated—called by his name, having recognized that it 'really' was he who was called by God, recognizes that he is a subject, a subject *of* God, a subject subjected to God." (1971, 179) The subjectivity here means that beside being created as a subject in God's image there is also the reverse pressure, recurrent in Judeo-Christian theology, to see God as needing mankind.

14　We might note a secular parallel from mathematics to this existence by fiat. One says in mathematics that an object—number, point, relation, etc.—with a given property 'exists' if its definition is consistent, if, in other words, it is not contradictory for it to have the property in question.

15　Thus the tale in the *Zohar* of God, who having created the letters and lived in their company for two thousand years, asked each letter to plead its case to be the instrument for the creation of the world, a competition won by the letter *beit* which (therefore) is the first letter of the first word of the *Torah*.

16　One should perhaps set Husserl's blindness to images and his silence about it in a certain context: in all likelihood he opposed diagrams as 'unrigorous' and indeed when he wrote his essay, a group of French mathematicians operating under the name Nicholas Boubaki was implementing a project, conceptualized (and known to Husserl) to rewrite the whole of mathematics in the formal language of first-order set theory. The result was thousands of pages of 'words' made from a mathematical alphabet of a dozen or so symbolic 'letters' and governed by a linear syntax, without a single diagram; an entire corpus, alphabetizing in the name of a Platonist-inspired structuralist program of foundational rigor, the richest, most elaborated trans-alphabetic—ideo- and diagrammatic—discourse yet created.

17　Of course, if, as Derrida insists, 'text' is enlarged to *ecriture* to include all forms of 'writing' which in turn is interpreted to mean any instance of the abstract category of 'spacing,' then his grammatological universalization of the text and monadic, theory-of-everything, master-term *différance* behind it will obliterate the distinction between visual spacing, here mathematical diagrams, and alphabetic spacing, here his own text which is a writing about Husserl's writing about mathematics, and so be indifferent to what the sustained, intentional but unexamined absence of the former might signify in the texts of Western philosophy.

18 The psychological status of Socrates's daemon, as well as the entire phenomenon of "voice hearers" and its modern pathologization in terms of aural hallucination and/or as a symptom of psychosis, is explored at length in Leudar and Thomas 2000.

19 There is a larger issue here of the link between money and writing relevant to Seaford's locating of pre-Socratic metaphysics, particularly its monism, in the monetization of Greek society. It is impossible to imagine a system of circulating currency and exchange in the absence of a developed practice of writing—if only to stamp gold coins and record transactions. Money, as Seaford notes, originated in Lydia in 540 BCE. But, as I indicate here, alphabetic writing is already present in Greece some two centuries earlier, both for purposes of memorialization and the marking of property in the form of inscriptions on objects. This being so there is no reason to refuse the idea that it was the institution and circulation of writing and not currency that laid the ground for abstract thought, metaphysical and otherwise, as well as being the source of the monadism Seaford identifies in Heraclitus's writings; a monadism which, as I indicated in the previous chapter, is deeply embedded within alphabetic writing. One can also mention a link between money and the first person pronoun. An example is the written 'I' which appears in the sixteenth century on banknotes whose unnaturalized precursor is in a sense the self-enunciating objects mentioned here, a transactional enunciation which I have discussed in relation to mathematical and pictorial forms of deixis. (Rotman 1987/1993)

20 Powell (2002) advances an extreme version of the latter, arguing that not only was the Greek alphabet *created* for the specific purpose of writing song and oral poetry, specifically to inscribe Homer's epic, but the feat was moreover accomplished by a single, historically identifiable individual.

21 "What's real Danny? Is reality TV real? Are confessions you read on the internet real? The words are real, *someone* wrote them, but beyond that the question doesn't even make sense. Who are you talking to on your cell phone? In the end you have no fucking idea. We're living in a supernatural world Danny. We're surrounded by ghosts." (Egan 2006, 130)

22 There are several ways of accounting for characteristics that ghosts might have that are compressed here. There is the evolution-based one of Boyer in terms of a violation of 'natural' categories; Talmer's approach from linguistics, where the ghost literalizes metaphors of movement used to describe spatial extension; and Taylor's cognitive science approach to the concept 'ghost' in terms of the theory of Blending.

23 This fate was anticipated and aided by an independent tendency within quantum physics toward the discrete and the quantized, culminating in Edward Fredkin's hypothesis of a 'finite nature' where "at some scale, space and time

are discrete," and the universe itself is theorized as a computer ticking itself forward a quantum of time at a time. For some of the background to physics' antagonism to infinitary procedures, see Rotman 2000, chapter 3.

24 A fictional exploration of what possessing the subjectivity of a quantum self might be like and some of the social effects it could encounter are the themes of Greg Egan's science fiction novel *Quarantine* (1992). See Hayles 2005 for an extensive reading of Egan's work.

25 A ghost fostered by technologies of the virtual would be under no necessity to be a monad (as opposed to a collective or multiplicity); or to be conscious in a human sense; or understandable as a god or goddess or within any of the categories we have hitherto imposed on (what we have taken to be) extra-human sentience; and under no necessity to know or love us or even notice humans as isolated, conscious individuals. A ghost or 'intelligence' or some form of sentience and/or agency emerging from networked computers is a much worked theme in science fiction literature of the last three decades. An early and paranoid example is John Varley's *Press Enter*, where the emergent agency eliminates humans who seek to come into contact with it. (Varley 1984)

26 Debray 2004 works out an extended and richly illustrated discussion and justification of the mediological approach to the birth and transmogrification of the Judeo-Christian divinity. A brief, nonspecialized review of this essay is given in Rotman 2005.

27 I give short shrift here to the important corporeal practices which surround and are seemingly essential to the status and dissemination of the religious text. Only by examining the numerous and interrelated embodied practices and habits associated with the uptake and continued inculcation of the holy word—ways of praying, chanting, laying of hands, fingering of iconic objects, pilgrimages, prostrating, baptizing, self-touching, fasting, singing, davenning, genuflecting, and so on—do the mediological effects of these shifts become properly evident.

References

Acocella, Joan. 1998. "The Politics of Hysteria." *New Yorker* 74 (7). April 6. 64–79.

Agamben, Giorgio. 1993. *Infancy and History: On the Destruction of Experience.* Trans. Liz Heron. London: Verso.

———. 2000. *Means without Ends: Notes on Politics.* Trans. Vincenzo Binetti and Cesare Casarino. Minneapolis: University of Minnesota Press.

Althusser, Louis. 1971. *Lenin and Philosophy and Other Essays.* Trans. Ben Brewster. London: Monthly Review Press.

Amaral, Júlio Rocha do, and Jorge Martins de Oliveira. 2005. "Limbic System: The Center of Emotions." *Mind and Behavior* 5. At http://www.cerebromente.org .br/n05/mente/limbic_i.htm. Accessed November 2005.

Appel, Kenneth, and Wolfgang Haken. 1989. *Every Planar Map is Four Colorable.* Contemporary Mathematics, Volume 98. Providence, R.I.: American Mathematical Society.

Armstrong, David F., William C. Stokoe, and Sherman E. Wilcox. 1995. *Gesture and the Nature of Language.* Cambridge: Cambridge University Press.

Artaud, Antonin. 1958. *The Theatre and Its Double.* Trans. Mary Caroline Richards. New York: Grove Press.

Bailey, James. 1992. "First We Reshape Our Computers, Then Our Computers Reshape Us: The Broader Intellectual Impact of Parallelism." *Daedalus* 121 (1). Winter. 67–86.

Baldwin, James Mark. 1895. "Consciousness and Evolution." *Science* 2 (34). August. 219–23.

———. 1902. *Development and Evolution.* New York: Macmillan.

Barglow, Raymond. 1994. *The Crisis of the Self in the Age of Information: Computers, Dolphins, and Dreams.* London: Routledge.

Barthes, Roland. 1975. *The Pleasure of the Text.* Trans. Richard Miller. New York: Hill and Wang.

———. 1977. *Image, Music, Text.* Selected and Trans. Stephen Heath. London: Fontana/Collins, 1977.

———. 1986. *The Rustle of Language.* Trans. Richard Howard. New York: Hill and Wang.

———. 1998. *The Semiotic Challenge.* Trans. Richard Howard. New York: Hill and Wang.

Batchen, Geoffrey. 2001. *Each Wild Idea: Writing, Photography, History*. Cambridge, Mass.: MIT Press.

Baudrillard, Jean. 1983. *Simulations*. Trans. Paul Foss, Paul Patton, and Philip Beitchman. New York: Semiotext(e).

Benveniste, Emile. 1971. *Problems in General Linguistics*. Trans. Mary Elizabeth Meek. Coral Gables, Fla.: University of Miami Press.

Boyer, Pascal. 2001. *Religion Explained: The Evolutionary Origins of Religious Thought*. New York: Basic Books.

Brooks, Rodney. 2002. *Flesh and Machines: How Robots Will Change Us*. New York: Pantheon.

Brown, Chappell. 2006. *Cell Computing Advances Simulation*. At http://eet.com/news/97/979news/cell.html. Accessed February 2006.

Bulwer, John. 1649. *Pathomyotomia, or a Dissection of the Affections of the Minde*. London: printed by W. W. for Humphrey Moseley.

———. 1664. *Chirologia: or the Naturall Language of the Hand. Composed of the Speaking Motions and Discoursing Gestures Thereof. Whereunto Is Added Chironomia: or The Art of Manuall Rhetoricke. Consisting of the Naturall Expressions Digested by Art in the Hand as the Chiefest Instrument of Eloquence*. London: printed by Thomas Harper for Henry Twyford.

Cassell, Justine, T. Bickmore, L. Campbell, H. Vilhjalmsson, and H. Yan. 2001. "More than Just a Pretty Face: Conversational Protocols and the Affordances of Embodiment." *Knowledge-Based Systems* 14 (1–2). March. 55–64.

Cavarero, Adriana. 2005. *For More than One Voice: Toward a Philosophy of Vocal Expression*. Trans. and introduced by Paul Kottman. Stanford: Stanford University Press.

Chadwick, Helen. 1989. *Enfleshings*. London: Secker and Warburg.

Chafe, Wallace. 1985. "Linguistic Differences Produced by Differences in Speaking and Writing." In David R. Olson et al., eds. *Literacy, Language and Learning: The Nature and Consequences of Reading and Writing*. Cambridge: Cambridge University Press.

Chandrasekaran, B. 1981. "Natural and Social System Metaphors for Distributed Problem Solving: Introduction to the Issue." In *IEEE Transactions on Systems, Man, and Cybernetics* 11. January. 1–5.

Changeux, Jean-Pierre, and Alain Connes. 1995. *Conversations on Mind, Matter, and Mathematics*. Ed. and trans. M. B. DeBoise. Princeton: Princeton University Press.

Châtelet, Gilles. 2000. *Figuring Space: Philosophy, Mathematics, and Physics*. Trans. Robert Shaw and Muriel Zagha. Dordrecht: Kluwer Academic Publishers.

Clark, Andy. 2003. *Natural-Born Cyborgs: Minds, Technologies, and the Future of Human Intelligence*. Oxford: Oxford University Press.

———. 2006. "Natural Born Cyborgs." *The Third Culture*. At http://www.edge.org/3rd_culture/clark/clark_index.html. Accessed April 2006.

Clarke, Bruce, and Linda Dalrymple Henderson. 2002. *From Energy to Information: Representation in Science and Technology, Art, and Literature*. Stanford: Stanford University Press.

Coleridge, Samuel Taylor. 1809. *The Friend*. Volume 1. Essay 3.

Comolli, Jean-Louis. 1980. "Machines of the Visible." In Teresa de Lauretis and Stephen Heath, eds. *The Cinematic Apparatus*. New York: St. Martin's Press. 121–42.

Corazza, Eros. 2004. *Reflecting the Mind: Indexicality and Quasi-indexicality*. Oxford: Oxford University Press.

Crary, Jonathan. 1988. "Modernizing Vision." In Hal Foster, ed. *Vision and Visuality*. Seattle: Bay Press.

Dantzig, Tobias. 1930/1985. *Number, the Language of Science: A Critical Survey Written for the Cultured Non-mathematician*. New York: Free Press.

Davis, Erik. 1998. *Techgnosis: Myth, Magic, and Mysticism in the Age of Information*. New York: Three Rivers Press.

Deacon, Terrence. 1997. *The Symbolic Species: The Co-evolution of Language and the Brain*. New York: Norton.

Debray, Regis. 2004. *God: An Itinerary*. Trans. Jeffrey Mehlman. London: Verso.

Dehaene, Stanislas. 1997. *The Number Sense*. Oxford: Oxford University Press.

de Kerckhove, Derrick. 1981. "A Theory of Greek Tragedy." *SubStance* 29. 23–36.

———. 2006. "Communication in Evolution: Social and Technological Transformation. An Interview with Derrick de Kerckhove, Director, McLuhan Program, Conducted by Álvaro Bermejo." In *The McLuhan Program in Culture and Technology*. At http://www.mcluhan.utoronto.ca/article_communication evolution.htm. Accessed April 2006.

Deleuze, Gilles. 1983. *Nietzsche and Philosophy*. Trans. Hugh Tomlinson. New York: Columbia University Press.

———. 1986. *Cinema I: The Movement Image*. Trans. Hugh Tomlinson and Barbara Habberjam. Minneapolis: University of Minnesota Press.

———. 1993. *The Fold: Leibniz and the Baroque*. Trans. Tom Conley. Minneapolis: University of Minnesota Press.

Deleuze, Gilles, and Félix Guattari. 1988. *A Thousand Plateaus*. Trans. Brian Massumi. London: Athlone Press.

Derrida, Jacques. 1976. *Of Grammatology*. Trans. Gayatri Spivak. Baltimore: Johns Hopkins University Press.

DiCenso, James. 1999. *The Other Freud*. Routledge: London.

Dixon, Joan Broadhurst, and Eric J. Cassidy, eds. 1998. *Virtual Futures: Cyberotics, Technology and Post-human Pragmatism*. London: Routledge.

Donald, Merlin. 1991. *Origins of the Modern Mind: Three Stages in the Evolution of Culture and Cognition*. Cambridge, Mass.: Harvard University Press.

———. 1998. "Mimesis and the Executive Suite: Missing Links in Language Evolution." In James R. Hurford et al., eds. *Approaches to the Evolution of Language: Social and Cognitive Bases*. Cambridge: Cambridge University Press. 44–67.

Edgerton, Samuel Y. 1980. "The Renaissance Artist as Quantifier." In Margaret Hagen, ed. *The Perception of Pictures*, Vol. 1. New York: Academic Press.

Efron, David. 1941/1972. *Gesture, Race and Culture*. The Hague: Mouton. First published as *Gesture and Environment*. New York: King's Crown Press.

Egan, Greg. 1992. *Quarantine*. New York: Harper.

Egan, Jennifer. 2006. *The Keep*. New York: Alfred A. Knopf.

Eisenstein, Elizabeth. 1983. *The Printing Press as an Agent of Change*. Cambridge: Cambridge University Press.

Elkins, James. 1995. "Art History and Images that Are Not Art." *Art Bulletin* 72 (4). December. 553–71.

———. 1999. *What Painting Is*. New York: Routledge.

Federman, Mark. 2005. "The Ephemeral Artefact: Visions of Cultural Experience." In McLuhan Program in Culture and Technology. At http://www.mcluhan.utoronto.ca/EphemeralArtefact.pdf. Accessed December 2005.

Feynman, Richard P. 1982. "Simulating Physics with Computers." *International Journal of Theoretical Physics* 21 (6–7). 467–88.

Finkelstein, Israel, and Neil Silberman. 2001. *The Bible Unearthed*. New York: Free Press.

Fischer, Steven. 2003. *A History of Reading*. London: Reakton Books.

Foster, Hal, ed. 1988. *Vision and Visuality*. Seattle: Bay Press.

Fredkin, Ed. 2001. *Introduction to Digital Philosophy*. At http://digitalphilosophy.org. Accessed April 2005.

Friedhoff, Richard Mark, and William Benzon. 1989. *Visualization: The Second Computer Revolution*. New York: Abrams.

Friedman, Richard. 1987. *Who Wrote the Bible?* San Francisco: Harper.

Furniss, Mauren. 2006. "Motion Capture." MIT *Communications Forum*. Cambridge, Mass.: MIT. At http://web.mit.edu/comm-forum/papers/furniss.html. Accessed February 2006.

Gabriel, Richard A. 2002. *Gods of Our Fathers: The Memory of Egypt in Judaism and Christianity*. Westport, Conn.: Greenwood Press.

Gigerenzer, Gerd. 2007. *Gut Feelings: The Intelligence of the Unconscious*. New York: Viking.

Gladwell, Malcolm. 2005. *Blink: The Power of Thinking without Thinking*. New York: Little, Brown and Company.

Goldreich, Oded. 1995. "Probabilistic Proof Systems." *Electronic Colloquium on Computational Complexity, Lecture Notes Series*. At http://www.eccc.uni-trier.de/eccc-local/ECCC-LectureNotes/goldreich/oded.html. Accessed April 2005.

Goody, Jack. 1977. *The Domestication of the Savage Mind*. Cambridge: Cambridge University Press.

Gordon, Avery F. 1997. *Ghostly Matters: Haunting and the Sociological Imagination*. Minneapolis: University of Minnesota Press.

Greenspan, Donald. 1973. *Discrete Models*. Reading, Mass.: Addison-Wesley.

Guattari, Félix. 1992. "Regimes, Pathways, Subjects." In Jonathan Crary and Sanford Kwinter, eds. *Incorporations*. New York: Zone. 16–35.

———. 1995. *Chaosmos: An Ethico-Aesthetic Paradigm*. Bloomington: Indiana University Press.

Hacking, Ian. 1995. *Rewriting the Soul: Multiple Personality and the Science of Memory*. Princeton: Princeton University Press.

Hansen, Mark. 2000. *Embodying Technesis: Technology beyond Writing*. Ann Arbor: University of Michigan Press.

———. 2004. *New Philosophy for New Media*. Cambridge, Mass.: MIT Press.

Hanson, Andrew J., Tamara Munzner, and George Francis. 1995. "Interactive Methods for Visualizable Geometry." At http://www.geom.umn.edu/~munzner/ieee94/ieee/ieee.html. May 2006.

Haraway, Donna. 2003. *The Companion Species Manifesto: Dogs, People, and Significant Otherness*. Chicago: Prickly Paradigm Press.

Harris, Roy. 1995. *Signs of Writing*. London, Routledge.

Havelock, Eric Alfred. 1963. *Preface to Plato*. Cambridge, Mass.: Belknap Press of Harvard University Press.

Hayles, N. Katherine. 1999. *How We Became Post-Human: Virtual Bodies in Cybernetics, Literature, and Informatics*. Chicago: University of Chicago Press.

———. 2005. *My Mother Was a Computer: Digital Subjects and Literary Texts*. Chicago: University of Chicago Press.

Hayward, Philip, ed. 1990. *Culture, Technology and Creativity in the Late Twentieth Century*. London: John Libby.

Hayward, Philip. 1990a. "Industrial Light and Magic." In Philip Hayward, ed. *Culture, Technology and Creativity in the Late Twentieth Century*. London: John Libby. 125–48.

Heller-Roazen, Daniel. 2002. "Speaking in Tongues." *Paragraph* 25 (2). 92–115.

———. 2005. *Echolalias: On the Forgetting of Language*. Cambridge, Mass.: MIT Press.

Hockney, David. 1993. *That's the Way I See It*. San Francisco: Chronicle Books.

Hornsby, Roy. 2001. "Virtual Reality, Cybernetics and Teledildonics in Computer Mediated Communication. (Or. . . . Will It Be Possible to Express Our Thoughts, Feelings, Emotions and Desires through an Electronic Communication System?)." At http://royby.com/cyber-culture/pages/cyberfutures.html. Accessed May 2005.

Hume, David. 1739–40/1951. *A Treatise of Human Nature*. London: Dent.

Hurford, James R., Michael Studdert-Kennedy, and Chris Knight, eds. 1998. *Approaches to the Evolution of Language: Social and Cognitive Bases*. Cambridge: Cambridge University Press.

Husserl, Edmund. 1970. *The Origin of Geometry*. In *The Crisis of European Sciences and Transcendental Phenomenology. An Introduction to Phenomenology*. Trans. David Carr. Evanston, Ill.: Northwestern University Press. 353–78.

Hutchins, Edwin. 1995. *Cognition in the Wild*. Cambridge, Mass.: MIT Press.

Ifrah, George. 1985. *From One to Zero: A Universal History of Numbers*. Trans. Lowell Blair. New York: Viking.

Illich, Ivan, and Barry Sanders. 1988. *A B C: The Alphabetization of the Popular Mind*. San Francisco: North Point Press.

Jakobson, Roman, and Morris Halle. 1971. *Fundamentals of Language*. The Hague: Mouton.

Kahn, Douglas. 1999. *Noise, Water, Meat*. Cambridge, Mass.: MIT Press.

Kaufmann, William J. III, and Larry L. Smarr. 1993. *Supercomputing and the Transformation of Science*. New York: Scientific American Library; distributed by W. H. Freeman.

Kendon, Adam. 1972. "Some Relationships between Body Motion and Speech." In Aron Wolfe Seigman and Benjamin Pope, eds. *Studies in Dyadic Communication*. New York: Pergamon Press. 177–210.

———. 1988. *Sign Languages of Aboriginal Australia: Cultural, Semiotic and Communicative Perspectives*. Cambridge: Cambridge University Press.

Kern, Stephen. 1983. *The Culture of Time and Space: 1880–1918*. Cambridge, Mass.: Harvard University Press.

Kipling, Rudyard. 1994. *Collected Stories*. New York: Alfred Knopf. Everyman's Library.

Kittler, Friedrich. 1992. *Discourse Networks: 1800/1900*. Stanford: Stanford University Press.

———. 1999. *Gramophone, Telephone, Typewriter*. Trans. Geoffrey Winthrop-Young and Michael Wutz. Stanford: Stanford University Press.

Knoespel, Kenneth. 2000. "Diagrammatic Writing and the Configuration of Space." In Gilles Châtelet. *Figuring Space: Philosophy, Mathematics, and Physics*. Trans. Robert Shaw and Muriel Zagha. Dordrecht: Kluwer Academic Publishers. ix–xxiii.

———. 2004. "Diagrammes, matérialité et cognition." *Theorie, littérature, enseignement* 22.143–63.

Kroker, Arthur, and Marylouise Kroker. 1996. *Hacking the Future: Stories for the Flesh-eating 90s*. New York: St. Martin's Press.

Kurzweil, Ray. 2005. *The Singularity is Near: When Humans Transcend Biology*. New York: Viking Penguin.

Lackoff, George, and Rafael E. Núñez. 2000. *Where Mathematics Comes From: How the Embodied Mind Brings Mathematics into Being*. New York: Basic Books.

Lambropoulos, Vassilis. 1993. *The Rise of Eurocentrism: Anatomy of Interpretation*. Princeton: Princeton University Press.

Landauer, Rolf. 1991. "Information Is Physical." *Physics Today*. May. 23–29.

Lane, Harlan 1984. *What the Mind Hears: A History of the Deaf*. New York: Random House.

Langer, Susan. 1951. *Philosophy in a New Key: A Study in the Symbolism of Reason, Rite, and Art*. Cambridge, Mass.: Harvard University Press.

Latour, Bruno. 1993. *We Have Never Been Modern*. Trans. Catherine Porter. Cambridge, Mass.: Harvard University Press.

Lecercle, Jean-Jacques. 2002. *Deleuze and Language*. New York: Palgrave Macmillan.

LeDoux, Joseph. 1994. "Emotion, Memory, and the Brain." *Scientific American* 5: 32–39.

———. 1996. *The Emotional Brain: The Mysterious Underpinning of Emotional Life*. New York: Simon and Schuster.

Leivant, Daniel, ed. 1995. *Logic and Computational Complexity. International Workshop, LCC '94, Indianapolis, IN, USA, October 13–16, 1994: Selected Papers*. New York: Springer.

Lenoir, Tim, and Xin Wei Sha. 2002. "Authorship and Surgery: The Shifting Ontology of the Virtual Surgeon." In Bruce Clarke and Linda Dalrymple Henderson, eds. *From Energy to Information: Representation in Science and Technology, Art, and Literature*. Stanford: Stanford University Press. 2002. 283–308

Leroi-Gourhan, André. 1993. *Gesture and Speech*. Trans. Anna Bostock Berger and introduction by Randall White. Cambridge, Mass.: MIT Press.

Leudar, Ivan, and Philip Thomas. 2000. *Voices of Reason, Voices of Insanity: Studies of Verbal Hallucination*. London: Routledge.

Levy, Pierre. 1997. "La virtualisation de l'intelligence et la constitution du sujet." At http://www.univ.paris8.fr/~hyperion/pierre/virt7. Accessed November 1997.

———. 1998. *Becoming Virtual*. Trans. Robert Bononno. New York: Plenum.

London, Barbara. 1996. "Video Spaces." *Performing Arts Journal* 18 (3). 14–19.

Lyotard, Jean-Francois. 1984. *The Postmodern Condition: A Report on Knowledge*. Manchester: Manchester University Press.

Maffeosoli, Michel. 1994. *The Time of the Tribes: The Decline of Individualism in Mass Society*. Trans. Don Smith. London: Sage Publications.

Manovich, Lev. 2001. *The Language of New Media*. Cambridge, Mass.: MIT Press.

Martin, Henri-Jean. 1994. *The History and Power of Writing*. Trans. Lydia Cochrane. Chicago: University of Chicago Press.

Martinec, Radan. 2004. "Gestures that Co-occur with Speech as a Systematic Resource: The Realization of Experiential Meanings in Indexes." *Social Semiotics* 14 (2). 193–213.

McClure, W. 1997. "Solution of the Robbins Problem." *Journal of Automatic Reasoning* 19 (3). 263–76.

McLuhan, Marshall. 1962. *The Gutenberg Galaxy: The Making of Typographic Man*. London: Routledge.

McNeill, David. 1992. *Hand and Mind: What Gestures Reveal about Thought*. Chicago: University of Chicago Press.

———. 2005. *Gesture and Thought*. Chicago: University of Chicago Press.

Merleau-Ponty, Maurice. 1962. *Phenomenology of Perception*. Trans. Colin Smith. London: Routledge.

Minsky, Marvin. 1987. *The Society of Mind*. London: Picador.

Mitchell, William J. 1994. *The Reconfigured Eye: Visual Truth in the Post-Photographic Era*. Cambridge, Mass.: MIT Press.

Mitchell, William T. 1984. "What is an Image?" *New Literary History* 15 (3). Spring. 503–37.

Murphie, Andrew, and John Potts. 2003. *Culture and Technology*. New York and Basinstoke: Palgrave Macmillan.

Nemirovsky, Ricardo, and Francesca Ferrara. 2004. *New Avenues for the Micro-analysis of Mathematics Learning: Connecting Talk, Gesture, and Eye Motion*. At http://www.terc.edu/mathofchange/EyeTracking/EyeTrackerPaper.pdf. Accessed May 2006.

Nietzsche, Friedrich. 1968. *The Will to Power*. Trans. Walter Kaufmann. New York: Vintage Books.

Núñez, Rafael E. 2006. "Do Real Numbers Really Move? Language, Thought, and Gesture: The Embodied Cognitive Foundations of Mathematics." In Reuben Hersh, ed. *18 Unconventional Essays on the Nature of Mathematics*. New York: Springer. 160–81.

Oakley, David A., ed. 1985. *Brain and Mind*. London: Methuen.

Oakley, David A., and Lesley Eames. 1985. "The Plurality of Consciousness." In David Oakley, ed. *Brain and Mind*. London: Methuen. 217–51.

Olson, David. 1994. *The World on Paper*. Cambridge: Cambridge University Press.

Olson, David R., Nancy Torrance, and Angela Hildyard, eds. 1985. *Literacy, Language and Learning: The Nature and Consequences of Reading and Writing*. Cambridge: Cambridge University Press.

Ong, Walter. 1982. *Orality and Literacy: The Technologizing of the Word*. London, Methuen.

Ouaknin, Marc-Alain. 1999. *Mysteries of the Alphabet: The Origins of Writing*. Trans. Josephine Bacon. New York: Abbeville Press.

Oxnam, Robert. 2005. *A Fractured Mind: My Life with Multiple Personality Disorder*. New York: Hyperion.

Papadimitriou, Christos H. 1994. *Computational Complexity*. Reading, Mass.: Addison-Wesley.

Phillips, Helen. 2006. "Known Unknowns." *New Scientist* 2582. December 16–22. 28–31.

Porush, David. 1998. "Telepathy: Alphabetic Consciousness and the Age of Cyborg Illiteracy." In Joan Broadhurst Dixon and Eric J. Cassidy, eds. *Virtual Futures: Cyberotics, Technology and Post-human Pragmatism*. London: Routledge. 45–64.

Powell, Barry. 1994. Review of Rosalind Thomas, *Literacy and Orality in Ancient Greece*. In *Electronic Antiquity* 1 (8). April. At http://scholar.lib.vt.edu/ejournals/ElAnt/V1N8/powell.html. Accessed May 2006.

———. 2002. *Writing and the Origins of Greek Literature*. Cambridge: Cambridge University Press.

Ragousi, Eirene. 2001. "The Hellenic Alphabet: Origins, Use, and Early Function." *Aristoriton*, Issue E014 (December). At http://phoenicia.org/alphabet controv.html. Accessed May 2006.

Rashevskii, Petr K. 1973. "On the Dogma of Natural Numbers." *Russian Mathematical Surveys* 28 (4). 143–48.

Reinhold, Arnold G. 1995. "P = NP Doesn't Affect Cryptography." Post to the sci.crypt,alt.security.pgp Newsgroup. 2 November. At http://www.world.std .com/~reinhold/p=np.txt. Accessed May 2006.

Richards, Robert. 1987. *Darwin and the Emergence of Evolutionary Theories of Mind and Behavior*. Chicago: University of Chicago Press.

Roco, Mihail. 2006. "Nanotechnology's Future." *Scientific American* 295 (2). 39.

Rotman, Brian. 1987/1993. *Signifying Nothing: The Semiotics of Zero*. London: Macmillan; Stanford: Stanford University Press.

———. 1993. *Ad Infinitum . . . The Ghost in Turing's Machine: Taking God Out of Mathematics and Putting the Body Back In*. Stanford: Stanford University Press.

———. 1998. "Response to Patrick Peccatte." Post of 27 August to Foundations of Mathematics List. At http://www.cs.nyu.edu/pipermail/fom/1998-August/002010.html. Accessed May 2006.

———. 2000. *Mathematics as Sign: Writing, Imagining, Counting*. Stanford: Stanford University Press.

———. 2000a. "Aura." *River City* 20 (2). 14.

———. 2002. "The Alphabetic Body." *Parallax* 8 (1). 92–104.

———. 2003. "Will the Digital Computer Transform Classical Mathematics?" *Philosophical Transactions, Royal Society*, London. Series A, 361. 1675–90.

———. 2005. "Monobeing." *London Review of Books* 27 (4). February. 17.

Sass, Louis. 1992. *Modernism and Madness: Insanity in the Light of Modern Art, Literature, and Thought*. New York: Basic Books.

Sazonov, Vladimir. 1995. "On Feasible Numbers." In Daniel Leivant, ed. *Logic and Computational Complexity. International Workshop, LCC '94, Indianapolis, IN, USA, October 13–16, 1994: Selected Papers*. New York: Springer. 30–51.

———. 1998. "Reply to Rotman." Post of 31 August on Foundations of Mathematics List. At http://www.cs.nyu.edu/pipermail/fom/1998-August/002031.html. Accessed May 2006.

Schwartz, Hillel. 1996. *The Culture of the Copy*. New York: Zone Books.

Seaford, Richard. 2004. *Money and the Early Greek Mind: Homer, Philosophy, Tragedy*. Cambridge: Cambridge University Press.

Shaviro, Steven. 1993. *The Cinematic Body*. Minneapolis: University of Minnesota Press.

————. 1995. *Truddi Chase*. At http://www.dhalgren.com/Doom/ch14.html. Accessed May 2006.

Sheets-Johnstone, Maxine. 2002. "Sensori-kinetic Understandings of Language: An Inquiry into Origins." *Evolution of Communication* 3 (2). 149–83.

Siegler, Robert S. 1996. *Emerging Minds: The Process of Change in Children's Thinking*. Oxford: Oxford University Press.

Sipser, Michael. 1992. "The History and Status of the P versus NP question." *Proceedings of the 24th Symposium on Theory of Computing*. May. New York: ACM Press. 603–18.

Snell, Bruno. 1960. *The Discovery of the Mind*. New York: Harpers.

Soloman, Ron. 1995. "On Finite Simple Groups and Their Classification." *Notices of the AMS* 42. February. 231–39.

Speer, Shari, R. G. Crowder, and L. M. Thomas. 1993. "Prosodic Structure and Sentence Recognition." *Journal of Memory and Language* 32 (3). 336–58.

Steigler, Bernard. 2001. "Hypostase, Phantasmes, Desincarnations." In Daniel Parrochin, ed. *Penser Les Reseaux*. Seyssel: Editions Champ Vallon. 136–48.

Stenning, Keith. 2002. *Seeing Reason: Image and Language in Learning to Think*. Oxford: Oxford University Press.

Stephens, Mitchell. 1998. *The Rise of the Image, the Fall of the Word*. New York: Oxford University Press.

Stevenson, Robert Louis. 1886/1979. *The Strange Case of Dr. Jekyll and Mr. Hyde*. London: Penguin Books.

Talmy, Leonard. 2000. *Toward a Cognitive Semantic*, 2 vols. Cambridge, Mass.: MIT Press.

Taylor, Mark, and Esa Saarinen. 1994. *Imagologies*. New York: Routledge.

Tedlock, Dennis. 1983. *The Spoken Word and the Work of Interpretation*. Philadelphia: University of Pennsylvania Press.

Thomas, Rosalind. 1992. *Literacy and Orality in Ancient Greece*. Cambridge: Cambridge University Press.

Turner, Mark. 2003. "The Ghost of Anyone's Father." *Shakespearean International Yearbook* 4. October. 72–97.

Varley, John. 1984 *Press Enter*. New York: Davis Publications.

Vygotsky, Lev. 1986. *Thought and Language*. Trans. and ed. by Alex Kozulin. Cambridge, Mass.: MIT Press.

Wheeler, J. H. 1988. "World as System Self-synthesized by Quantum Networking." *IBM Journal of Research and Development* 32 (1). 4–15.

Wilbur, Richard. 1988. "An Event." In *New and Collected Poems*. New York: Harcourt.

Willis, Anne-Marie. 1990. "Digitisation and the Living Death of Photography." In Philip Hayward, ed. *Culture, Technology and Creativity in the Late Twentieth Century*. London: John Libby. 197–208.

Wilson, Elizabeth. 2004. *Psychosomatic: Feminism and the Neurological Body*. Durham: Duke University Press.

Wise, M. Norton, ed. 2004. *Growing Explanations: Historical Perspectives on Recent Science*. Durham: Duke University Press.

Wolfram, Stephen. 2002. *A New Kind of Science*. Champaign, Ill.: Wolfram Media.

Zumthor, Paul. 1990. *Oral Poetry: An Introduction*. Trans. Kathryn Murphy-Judy. Minneapolis: University of Minnesota Press.

Index

Art, 19; aura of, 96
Artaud, Antonin, 4, 39, 48, 49, 104;
   The Theatre and Its Double, 39
Artificial intelligence, 91
Artificial speech synthesis, 23
Asymptotic growth, 71–75
Attitude of the speaker, xxxiii, 26, 27,
   126
"Aura" (prose poem), xxxviii
Australopithecine, xvii
Auto-hapticity, 38
Avatars, 8, 46, 134; sex between, 47

Babel, 16
Bacon, Francis, 16
Bailey, James, 90
Baldwin, James Mark, xviii
Baldwinian evolution, 115
Barglow, Raymond, v, 81
Barthes, Roland, 5, 23, 27, 126, 145n12
Baudrillard, Jean, 95
Becoming, iv, 103
Becoming beside ourselves, 103, 104,
   105, 134
Benjamin, Walter, 95
Benveniste, Emile, 107–8
Benvenistian linguistics, 108
Benzon, William, and Richard Mark
   Friedhoff, 66
Beside, iv, 103
Blind people, 21
Body, the, 41, 53; alphabetic writ-
   ing and, 41; as exogenous, 133; as
   heterotopic, 134; as increasingly
   readable, 134; obsolescence of, 46;
   as transparent, 133
Body practices, 108; in religion, 150n27
Boolean logic, 69
Borges, Jorge Luis, 94
Bounds, 70, 71, 72, 73
Bourbaki, Nicolas, 66
Boyer, Pascal, 131

Brain-machine interfaces, 139n1
Brain size, xx, xxvi
Brooks, Rodney, x
Brosses, Charles de, 16
Brushstrokes, 42–45
Bulwer, John, 16, 17; *Chirologia*, 16,
   140n4; *Panthomyotomia*, 16

Calculus, 65, 89; linearity of, 144n6
Calvino, Italo, 121
Cantor, Georg, xxxii, 132; infinite
   arithmetic of, 87
Cassell, Justine T., et al., 48
Cavarero, Adriana, 121–22, 147n11
Cellular automata, 62, 89
Chadwick, Helen, xxviii, 95
Chafe, Wallace, 22
Chandrasekaran, B., 89–90
Chaos theory, 62
Châtelet, Gilles, 35–38, 45, 141n3
Children's thinking, 34
Chinese orthography, 39
Chinese writing, 13, 94, 124
Christianity, 136
Church, Alonzo, xv
City-states, 13
Clark, Andy, xiii, 1, 8
Classical music, 86–88
Co-evolution, x, xvi, xvii; of mathe-
   matics and machines, 58–59, 62
Cognitive ethnography, 91
Cognitive science, xvi
Cognitivism, 91
Coleridge, Samuel Taylor, 107
Commandments, 120, 124
Commoli, Jean-Louis, 96, 143n1
Complexity theory, 71, 73
Computability, xv
Computational sciences, xiv, xvi
Computer mouse, 44, 98, 110
Condillac, Étienne Bonnot de, 16
Connes, Alain, 38

Consciousness: postmodern, 100; of the self in Deacon, xxii; through writing, 94

Co-presence, 83, 98, 99, 104

Corazza, Eros, 115

Corporeal writing, xxiv

Cortical function, 90, 128

Cortical origins, 24–25

Counting, 87, 102, 131; as creation, 76; metaphysics and, xxxii; real vs. imagined, 76; with whole numbers, 85

Crary, Jonathan, 97

Cryptography, 62, 69, 72, 73, 74; achievability of decryption and, 71; pragmatic concerns and, 73

Culture, 51–52

Cyborgs, xiii, xvii, 1

Dantzig, Tobias, 85, 86

Davis, Erik, 117

Deacon, Terence, xxii, xxv, 23, 24, 30, 52, 109, 114, 115, 116, 118, 139n6; on symbolic web, 114. *See also* Deacon, Terence, and Merlin Donald

Deacon, Terence, and Merlin Donald, xvii, xx; *Homo Symbolicus*, xvii, xx; on language as life form, xviii

Deafness, 16

Debray, Regis, 120, 136, 150n26

Deconstructionist discourse, 42

Dehaene, Stanislaus, xxiii–xxiv, 52, 102

de Kerckhove, Derrick, 8, 126, 134, 135

Deleuze, Gilles, 53, 134, 140n1, 145–46n2. *See also* Deleuze, Gilles, and Félix Guattari

Deleuze, Gilles, and Félix Guattari, iv, xiii, xiv, xxiii, xxvi, 104, 119

De-prosodized speech, 31

Derrida, Jacques, xiv, xxiv, 124, 139n2, 148–49n17; *Of Grammatology*, xv;

the subject in, 139n2; on writing as graphism, xxiv

Descartes, René, 128

Deus ex machina, 116–19

Dewey classification, 94

Diagrams, 33–40, 58, 124, 125, 141n14, 142n2, 143n3, 148–49n17; absence from "The Origin of Geometry," 125, 148n16; as allusions, 37; as epiphenomenal, 66; as 'frozen gestures,' 37, 108; as more than representations, 37; in painting, 45; symbols in, 42

Diaries, 94

Digital binary code, 2

Digital computer, 59; antagonism of, to the analog, 76; effect of, on mathematicians, 61; effect of, on mathematics 59, 75–77; recursive capacity of, 75–76; testing and solution and, 143n2

Digitally produced images, 3, 40, 93. *See also* Images

Digital mathematics, 132

Digital media, 144n8; consequences of, for monotheism, 136

Discretization, 65–66

Discursive communication, 84

Distributed cognition, xv

"Dogma of Natural Numbers" (Rashevskii), 75

Donald, Merlin, xxvi, 51, 84, 103, 115, 139n7, 145n11, 146n3. *See also* Deacon, Terence, and Merlin Donald

Doyle, Arthur Conan, 117

Eames, Lesley, and David A. Oakley, 101

Edgerton, Samuel Y., 95

Edison, Thomas, 117

Efron, David, 18

Egan, Greg, 150n24

Gesture and Speech (Leroi-Gourhan), 39

Gesturo-haptic, the, 50; difference between textual writing and, 51; in theatre, 49; as 'visual notation,' 50

Gesturo-haptic body, 50, 109

Gesturo-haptic mediation, 51, 53

Gesturo-haptic writing, 39–54; consequences of, for writing and speech, 47; exo-textuality of, 50; gesturo-haptic medium and, 47, 109; possibilities of, 54

Gesturology, 4, 16, 50

Ghost effect, xiii, 7, 8, 107–37; from human speech, 112. *See also* God; Infinity; Mind

Ghostliness, bio-linguistic, 113–16

Ghostliness, biology of, 116

"Ghost of Anyone's Father, The," 113

Ghosts, 107; communicational media and, 116; definition of, 113; fictive, 131; medium-specificity of, 113; network-induced, 135; postmodernness of, 113; types of, 149n22; the virtual and, 150n25

GIS (Graphic Information System), 98

GIS map, xxvii, xxviii; Google Earth, xxviii; parallel and serial seeing and, xxviii

Globalization as planetary network, 135

GNU, 123

God, 8, 9, 54, 112, 116, 118, 130, 131, 132, 136, 148n13; alphabetic, 122; freedom from judgment of, 104; as 'I-effect,' 130; inside and outside of mathematics, xxxii; invisible, 110, 119; mathematics and, xxxii; as media-effect of the alphabet, xxxiii; obsolescence of, 136; shared origins of, with the 'mind,' xxxii; spoken word of, 26; writing and, 7, 124. *See also Jahweh*

Gödel, Kurt, xv

God particle, 137

Goldreich, Oded, 64

Gordon, Avery F., 113

Gorgian rhetoric, 3, 126

GPS, 8

Grammar, 4, 28, 50, 114

Graphic User Interface, 44, 98

Graph theory, 62, 69

Greek/Hellenic alphabet, 125, 128–29, 135, 140n2, 149n20

Greeks, 13–14, 118

Greenspan, Donald, 65

Guattari, Félix: on intimacy of technology and subjectivity 5, 82. *See also* Deleuze, Gilles, and Félix Guattari

H1 and H2 regions, 73

Hacking, Ian, 100, 145n15

Haken, Wolfgang, and Kenneth Appel, 64

Halle, Morris, and Roman Jakobson, 86

Hamlet, 23, 113

Hand, 39

Handshake, 51

Handwriting, 25

Hansen, Mark, xvii, 6

Hanson, Andrew J., 66

Haptic feedback, xvi; virtual auto-hapticity and, xxiv

Haraway, Donna, xvii; "Companion Species Manifesto," xvii

Havelock, Eric, 127, 143n1

Hayles, Katherine, xi, xii, xvii, 144n8, 145n17, 150n24

Hebrew language, 14, 110, 122

Hebrews, 118

Hegelian spiral, 84, 93

Heisenberg, Werner, 86

Heller-Roazen, Daniel, 124, 129

Heraclitus, 127, 149n19

Hilbert, David, xv

History, 39
Hockney, David, 98–99
Homer, 126, 149n20
Horizon effect, 87
Hornsby, Roy, 47
How We Became Posthuman (Hayles), xi
HTML, 40
Hugo, Victor, 93; in Ouaknin, 13
Human assemblages, xiii, xiv
Human cognition, flexibility of, xxii
'Human computers,' 90
Humanities, use of empirical data in, xxiii
Human mind, 1
Human nature, 103
Hume, David, 101–2
Husserl, Edmund, 125, 148n16, 148–49n17; Derrida on *The Origin of Geometry* and, 125; lack of diagrams in, 125; *The Origin of Geometry*, 125
Hutchins, Edwin, 90

I, xxxiii, 4; bleeding outward of, 99–100; evolution of language and, 109; ghost spirit of, 118; Greek, 125–30; "I am the Lord Thy God" and, 7, 24; immersive or gesturo-haptic, 8; Jewish, 119–25; lack of separation between inner and outer, 102; as medium-specific, 109; networked, 8, 133–37; plural, 104, 134; property and, 129, 149n19; reading, 135; as reading/writing agent, 4; shift away from the monoidal and, 83, 92, 103; spoken and written, 108, 110, 128; third, 8; in the *Torah*, 122; virtual, xxxiv; as Western 'Me,' 104; writing, 7, 31, 126; written vs. signed, xxxiii
I / Self / Other, 99–105
I, The, 107–10; linguistic character of, 107
Icon, xx, xxi, xxii, 3, 114, 115
Iconophobia, 40, 83

Idea units, 22
Ideograms. *See* Diagrams
'I-er,' 122, 123, 128, 129
Illich, Ivan, and Barry Sanders, 2, 14, 28, 94
Illocution, 26
Imaged images, 97–98
Imaged subjects, 93–99
Images, xvi, xxiii, xxiv, xxvi, xxvii, xxviii, 2, 3, 42, 43; digital or post-photographic, xxviii, 3, 40, 93; flat, 44; imaged, 97–98; instrumental; 95; in Mignone, 39; on-screen, 66; as parallel, 83; singularity or multiplicity of, 97; tension of, with words, 83–84; usurpation of alphabetic writing by, 40; visuo-kinetic, 22
Imaginary money, xv, xxxi
Imaging, 93–99
Imagology, 93
Imagologies (Taylor and Saarinen), 93
Index, xx, xxi, xxii, 21, 24, 114
Infants, xviii, 49, 115, 121
Infinite Mathematical Agent, 110, 130–33
Infinite sets, 86–88, 132
Infinity, 7, 68–77, 112, 130, 132; algorithms and, 71; computer science and, 68–75; displacement of, in mathematics, 62, 67; as ideal, 77; mathematical Person and, 77; opposition of computers to, 76; physics and, 65
Information, as embodied, xv
Installation art, 44
Instrumental Transcommunication (ITC), 117, 146n6
Instrumental view of technologies, 5–6
Interface design, 48
Integers, xxxii, 68, 74
Intermediation, xii
Internet, ix, xxvi, 46, 89; as tactile, 134

Materiality (*continued*)
37; of media, xiii, xxv, 6; of paint,
43; of technology 5, 103
Mathematical agent, xv, xxxi, 110,
130–33
Mathematical infinite, subjectivities
and, xxxi
Mathematician, 59, 60, 61, 63, 64, 67,
130
Mathematics: classical, 74, 75, 77;
computers and, 57–77; embodiment of, 33, 36–38; empirical (vs.
classical), 68, 74; images in, xvi; as
interplay of idea, symbol, and procedure, 59–60; as near-experimental
discipline, xvi; "new science" of, 68;
Person, Subject, and Agent in, 60–
61, 64, 67, 76, 130; proofs in, 60–63;
pure, 61; rigor of, 77; semiotics and,
xiv; simulation and, 57–58; technol-
ogization of, 58–62. *See also* Mathematician
McClure, W., 63
McLuhan, Marshall, 54, 127, 143n1
McNeill, David, 18, 22, 139n7, 141n8
Meaning, sameness of, 19
Measurement problem in quantum
physics, 86
Mechanical reproduction, 96
Media, ix, xii, xiii, xvii, xix, xxiii, 5,
7, 20, 41, 57, 102, 104, 107, 109, 112,
134–36, 143n1, 144n8; capture, 42;
diminishing returns of, 135; ecologies of, 8, 81; ghosts of, 113, 116,
117; memory, xxvi; network(ed),
xxxiii, 2, 5, 133; notational, 42; philosophers of, xiv, 93; regimes of,
xvi, xxvi; thought and, xxv; written,
53
Memory, 52, 84, 100, 133; conceptual,
85; death and, 145n5; episodic vs.
procedural, 85; evolution of cognition and, 145n11; viewing and, 98

Merleau-Ponty, Maurice, 34; triangle
of, 34–35
Metacommunication, 20
Metaphors, 42; in mathematics, 37–38
Metaphysics, Western, xv, xxiv, xxxi,
7, 92, 94, 137; alphabetic writing
and, 2, 110; in Derrida, 139n2; of the
infinite, xxxii; metaphysical thought
in mathematics and, 38; ontotheological, 9; Platonic, 38; I in, 99; the
voice and, 147n11
Metonymy and synecdoche, 42
Midbrain, xxv, 24, 30, 31, 128
Mignone, Christof, 39; "The Prestidigitator: A Manual," 39
Mimesis, 27, 28, 43
Mimicry, through media, 43
Mind, 7, 26, 31, 110, 112, 118, 127, 130,
136. See also *Nous*; Psyche
Mind-body dualism: in preference for
writing over speech, 30–31; separa-
tion of midbrain and neocortex and,
xxv; sign language and, 17
Minsky, Marvin, 101
Mitchell, William, 83, 96; on "onto-
logical aneurism," 96
Monadology, 94
Monetization of Greek society, 127,
149n19
Monotheisms, 119; consequences of
digital media for, 136; monolatry
vs., 120; text-based, 40; Western, 54
Moravec, Hans, ix
Morgan, Steve, 72
Morphology, 114
Morse, Samuel F. B., 116
Moses, 118–19, 122, 124, 147n9; interpellation and, 148n13
Motion capture technology, 4, 43, 45,
110, 134; communicative potentials
of, 47; decontextualization and, 46,
52; digital nature of, 45; muteness
and, 49; other names for, 142n7;

Starlings, 89
Steigler, Bernhard, 135
Stellarc, 46; transposed corporeality of, 46
Stevenson, Robert Louis, 81, 101
Strange Case of Dr. Jekyll and Mr. Hyde, The (Stevenson), 81
Stroking, 45, 46, 47
Subject, xxvi, xiii, xiv, 5, 83, 100–102, 104, 139n2; Althusser and, 148n13; analog and digital, xii; bio-technic, 53, computing, 88–92, ghostly, xvi, 'I'-saying, 108; imaged, 93–99; as linguistic abstraction, 108; mathematical, 60–61, 64, 67, 130; parallelist, 40; post-human, xxiv; speaking, 50; viewing, xxvii; virtual, 111
Subject formation, xii, xiii, xx, xxi
Subjectivity, xii, xiii, xiv, xxxi, xxxiii, 15, 111, 148n13; through the gesturohaptic, 51, 52; human mind and, 1; through images, 40; outside of language and signification, 82, 108; as medium-specific, 109; parallel computing and, 134; Person as, 60; pre-linguistic, 115; reference to I and, 108; quantum, 150n24; secondary, xv; techno-apparatuses and, 82; technologized, 51–54; technology and, 5; theatre and, 49; visual, 95
Superposition, 85, 134
Surveillance, 8
Svenbro, Jesper, 129
Syllables, 86; as governed by neocortex, 29
Symbolic communication: in Deacon, 24; in Deacon and Donald, xvii, xviii; as extrabiological, xxiii
Symbolic reference, xix, 114, 115, 139n6; as efficient, xx; as interpreter-independent, xxii; as shared, xxii; as specifically human, xx; subject formation and, xx; as virtual, xx, xxi

Symbolic relationships, xix, xxi; between tokens, xxi
Symbols, xviii, xix, xx, xxi, xxii, 51, 59, 114, 116, 132, 139n5; in mathematics, 60, 124; musical, 42; symbol-processing brain, xxvii, 91; visual, 19
Syntax, 18, 28, 41, 42, 63, 75, 94, 95, 114, 140n5
Synthetic assemblage, 52, 102

Talmud, 13; interpretation of, 26; scholars of, 124
Talmy, Leonard, 131
Taylor, Mark, and Esa Saarinen, 93
Technesis, 6
Technics, xiii
Technoid subjects, 81–83
Technologized mathematics, 57–77
Technologized subjectivity, 51–54
Technology, ix, xvii, xix, xxv, 1, 2, 4, 39, 41, 46, 57, 58, 65, 73, 81, 88, 102, 111, 112, 117, 133; corporeal axiomatic of, 51; embodiment of, 6; materiality of, 4; as more than discursive process, 6; multitasking and, xxvii; neurological alterations of, 53; as prosthesis, xiii, 5, 53; psyche in, 81, 83; social and cognitive consequences of, 1; subject and, xxxv, 5, 51–54; time and space and, 53
Tele-dildonics, 47
Telegraphs, 7, 116; the dead and, 117; writing and, 117
Telephone, 6, 21, 117
Tele-surgery, 47
Television, 2, 93, 95, 117; picture in picture, 98
Texts, xxvi, 3, 4, 6, 26; affect in 27; electronic, 28; outside of, 124; post-human machine and, xii; prosody of, 110; purely textual entities and, 29
Textual brains, 29–31

Writing (*continued*)

over body in, 31; as re-mediation, 6; as spacing, 42; supernatural agency and, 14; virtual entities and, 6

Writing speech, 25–28; elimination of embodied gestures in, 25–27; reinvigorating speech and, 147n8; rescuing speech and, 122–23, temporal dimension of, 25, 118

Zazen, 103

Zero, xiv, xxxi, xxxii; as absence and origin, xxxi; consequences of, xiv–xv

0s and 1s sequences, 67

Zumthor, Paul, 27, 140n3

Brian Rotman is a professor of comparative studies
at Ohio State University. He is the author of several
books, including *Mathematics as Sign: Writing,
Imagining, Counting* and *Signifying Nothing: The
Semiotics of Zero*.

Library of Congress Cataloging-in-Publication Data
Rotman, B. (Brian)
Becoming beside ourselves : the alphabet, ghosts,
and distributed human being / Brian Rotman ;
foreword by Timothy Lenoir.
p. cm.
Includes bibliographical references and index.
ISBN-13: 978-0-8223-4183-3 (cloth : alk. paper)
ISBN-13: 978-0-8223-4200-7 (pbk. : alk. paper)
1. Science—Philosophy. I. Title.
Q175.R5655 2008
501—dc22   2007044861

Printed and bound by CPI Group (UK) Ltd, Croydon, CR0 4YY

27/10/2024

14580225-0004